Revaluing Coastal Fisheries

Alexander Dobeson

Revaluing Coastal Fisheries

How Small Boats Navigate New Markets and Technology

Alexander Dobeson
Uppsala University
Uppsala, Uppsala Län, Sweden

ISBN 978-3-030-05086-3   ISBN 978-3-030-05087-0   (eBook)
https://doi.org/10.1007/978-3-030-05087-0

This Palgrave Macmillan imprint is published by the registered company Springer Nature Switzerland AG
The registered company address is: Gewerbestrasse 11, 6330 Cham, Switzerland

Preface

While setting sail as a PhD student in the Department of Sociology at Uppsala University in 2011, my scholarly interest was mainly dedicated to the increasing role of markets in society. At the time, the consequences of markets and quasi-markets on public institutions such as health care, higher education and science had been widely studied. Moreover, the urban worlds of industrial, aesthetic and financial markets had become a gold mine for economic sociologists to formulate a comprehensive empirical critique of the dominating neoclassical market model. While reading more broadly into the economic sociology literature on markets and property rights, however, I soon noticed that Classics such as Karl Marx, Max Weber and Karl Polanyi were spending a notable amount of pages on the issue of how 'nature' is organised and distributed as important pillar of the modern market economy. To my surprise, however, this crucial aspect seemed rather neglected in the more contemporary economic sociology literature on markets, networks and organisations. Instead, issues around natural resources were mostly dealt with in other fields such as economic geography or environmental sociology. Having had no previous relation with fisheries other than recreational, it more or less came to me as a coincidence that I came across an article about the privatisation of fish stocks around a small island state in the middle of the North Atlantic ocean. Although traces of market-based fishery management could already be found in other fisheries, the case of the privatisation of the Icelandic marine commons seemed fairly extreme and was even said to have played its part in the country's financial meltdown in 2008. While all of this sounded very exotic and far away from the urban worlds studied at my department, I simply could not stop wondering what all the fuss around these so-called Individual Transferable Quotas (ITQs) was about. All of a sudden, I was motivated to finish my coursework as fast as possible and soon found myself boarding a plane bound to Keflavík airport at the beginning of April 2012.

Arriving for the first time in this estranged moon-like landscape, I did not have a clue how to study or even locate the elusive ITQs. For this reason, I decided to contact some colleagues at the University of Iceland in Reykjavík, most notably Gislí Pálsson whose early anthropological works sparked my interest in fisheries. To my surprise, however, many social scientists with a critical perspective on ITQs seemed to have grown rather frustrated with the subject. The same seemed to hold true for most of the locals who seemed to be tired of fierce and often emotional debates about the ownership of the nation's most valuable resource. As a young Icelandic woman put it: 'If you want to ruin a party in Iceland, ask everyone about their opinion on the quota system'. For many inhabitants of the capital region, the ITQ system seemed to epitomise everything that was wrong with the neoliberal course of the pre-crisis government and financial elite. Nevertheless, most people I met throughout my stay encouraged my project and welcomed my view on this highly emotional and politicised subject. In fact, I would learn much later that the outsider's perspective is in fact much appreciated in Icelandic culture, as reflected in an old Icelandic proverb that can roughly be translated to 'Keen is the eye of the guest' (Glöggt er gests augað).

Intrigued by the case, I decided to go on a few field trips to the coastal communities of the southwest. What I then saw, however, puzzled me even more. Contrary to all the talk about capitalist expansionism and capital concentration I noticed a considerable number of rather small and very modern looking fishing vessels in the harbours of many coastal communities. This however did not at all seem to fit the bill of all the talk about privatisation, markets and financial capitalism I was confronted with in post-crisis Iceland. After reporting my observations to some of my colleagues, it turned out that nobody really knew what was going on there without falling back to old-fashioned scripts. While some saw coastal fisheries as a sign of rural resilience against market orthodoxy and economies of scale, others seemed to see these small industries simply as a relic of the past that was about to vanish eventually. I was however not really convinced by either of these views and decided I had to find out for myself what these 'small boats' were all about. While I had to return to Sweden for my departmental duties, I did not take long and I returned to Iceland in the second half of the summer, although this time with transfer to a much smaller plane that would land right in between mountains on a short landing strip in the middle of an awe-inspiring fjord in the outer northwest known for sheltering the largest fleet of small boats in the country. It was from this moment on that I felt I had finally arrived in the world of contemporary coastal fisheries.

Enduring a fair amount of storms in the rural periphery of the North Atlantic did not only make me realise what it means to live and work in an often rough and unpredictable environment but also understand the true strength of the ethnographic method compared to other conventional, or for the lack of a better word: more 'rigid' methodological approaches. Being in the field enabled me to capture a first-hand experience of the local dynamics and practices that would have gone unnoticed and hardly been reflected in any survey-study or statistical dataset. Hanging out at the docks and fish auctions on a daily basis, going fishing trips with the locals engaging in community life revealed a different, more pragmatic side of the story than conventional theories of social organisation and capitalism suggest. It soon became clear to me that new markets and property rights do not simply 'intrude' and 'destroy' allegedly less alienated forms of community life. Thus, the aim of this book is to show that the consequences of neoliberal reforms are much more far-reaching than the one-sided story of market vs community suggests, as they revalue the material culture of the rural periphery as such.

The insights into this new culture of *liberal* rural capitalism would have been impossible without the support of all the people that gave me hospitality and support even during the coldest and windiest Icelandic summers and winters. In particular, I have to thank Claudia Matzdorf and my other roomies at the 'shrimp house' in Ísafjörður, my fishing companion from Engidalur and all other rural dwellers, fishers and fish workers I met during my time in the fishing communities of the Westfjords and the rest of Iceland.

My academic thoughts and ideas, however, were not exclusively developed in the field, nor did they simply come out of the blue. Like fishing, the development of new ideas is a social endeavour, in which skills and knowledge are passed on by masters to their apprentices. For this reason, I would like to thank Patrik Aspers and Vessela Misheva for their expertise and enduring support of this project. Furthermore, I am deeply grateful to Karin Knorr Cetina for inviting me to engage in further discussions on markets and technology and giving me time to work on the manuscript during my stay at the University of Chicago during the autumn of 2014. Moreover, the book would not be the same without the critical comments from my colleagues, most notably Kristin Asdal, Jahn Petter Johnsen, Mats Franzén, Árni Sverisson, Michael Allvin, Marie Sépulchre and Tobias Olofsson during different stages of the writing process. I thank you all equally for the work and support you have invested in improving the manuscript during my journey to this book.

The development of new ideas, however, would not be half the fun without the rest of the crew. For this reason, special thanks goes to the Uppsala Lab of Economic Sociology, namely Petter Bengtsson, Edvin Sandström, Ellinor Anderberg, Jonas Bååth, Carl Sandberg, Henrik Fürst, Elena Bogdanova, Dominik Döllinger, Ugo Corte, Clara Iversen, Alison Gerber, Sebastian Kohl, Sonia Köllner, Hannes Landén and Tom Chabosseau. The same goes for my other friends and colleagues at Uppsala University, most importantly Siavash Alimadadi, Mikael Svensson, Ylva Nettelblad, Philip Creswell, María Langa, Christoffer Berg, Nicklas Neumann and Inti José Lammi. Finally, I would like to thank Jeremy Phillipson for his kindness and inviting me to the Centre for Rural Economy at Newcastle University

(UK) in the autumn of 2017 where I got the chance to further develop and refine my ideas in exchange with the field of rural sociology. I would also like to thank Sally Shortall, Ruth McAreavey and Menelaos Gkartzios for their immeasurable hospitality and kindness during my stay in the Toon.

Some of the chapters are partly based on earlier versions that have appeared elsewhere. Parts of Chapters 3 and 8 appear in 'Economising the Rural. How New Markets and Property Rights Transform Rural Economies', *Sociologia Ruralis*, Vol. 58 (4), pp. 886–908 (with friendly permission from Wiley Publishers). Chapter 6 reproduces some material that has appeared in 'The Wrong Fish: Manoeuvering the Boundaries of Market-Based Resource Management', *Journal of Cultural Economy*, 2018, Vol. 11 (2), pp. 110–124 and parts of Chapter 7 in 'Scopic Valuations: How Digital Tracking Technologies Shape Economic Value', *Economy and Society*, 2016, Vol. 45 (3–4), pp. 454–478, both with friendly permission from Taylor and Francis. I thank the publishers for their kindness and support! It goes without saying that this book would have not been possible without the generous support of the ERC-funded project 'Convaluation: valuation and evaluation in- and outside the economy' (Starting Grant ERC 263699-CEV).

Last but not least, I could have not managed to write this book without all of my friends outside of academia who helped me navigating the highs and lows of becoming a scholar and most importantly, keeping me grounded. Thank you for bearing with me!

I dedicate this book to my parents Birgit and George Dobeson who have always supported me no matter what. Thanks for everything!

Uppsala, Sweden   Alexander Dobeson
October 2018

The original version of the book was revised: Corrections were made in the preface and various chapters of the book. The correction to the book is available at https://doi.org/10.1007/978-3-030-05087-0_10

Contents

List of Figures

1

Introduction

> *Sooner shall all the hummocks on Summerhouses*
> *land hop up to heaven and all the bogs sink down*
> *to bottomless bloody hell than I shall renounce my*
> *independence and my rights as a man.*
>
> Bjartur of Summerhouses of Halldór Laxness'
> *Independent People* (1946)

Bjartur's Metamorphosis

Some place far away from the night crawlers and bohemia of Reykjavík's infamous nightlife scene in a rural Icelandic fishing community, situated in the remote rural Westfjords. After a restless winter's night just below the Arctic Circle our main character—we shall call him Bjartur ('the bright') after the tragic hero of Summerhouses from Halldór Laxness' epic novel[1]—decides to get up around 3:30 a.m. and evaluate whether

---

[1] Halldór Laxness' novel *Independent People* (1946) (Icelandic original: *Sjálfstætt fólk,* 1934–1935) deals with the life and hardships of the peasant farmer Bjartur of Summerhouses who is struggling against the forces of nature, economic turmoil and the socio-political pressures with a desperate will to live an independent life in the countryside at the cost of his family, ending up debt-bound and financially devastated. Laxness remains the only Icelandic winner of the Nobel prize for literature (1955).

© The Author(s) 2019
A. Dobeson, *Revaluing Coastal Fisheries,*
https://doi.org/10.1007/978-3-030-05087-0_1

or not the weather conditions will finally allow him to follow his occupation as an independent fisher.[2] For the past fourteen days, strong westerly winds up to 30 metres per second have produced giant swells rolling into the fjord, making it virtually impossible to put out the hand-baited longlines on his small ten metre-long coastal fishing vessel. The struggle with the weather is nothing new for Bjartur who grew up in the harsh Arctic environment. The villagers are continuously confronted with the omnipotence of nature and the constant potential for disaster louring from the ocean and the mountains that embrace the fjord. Especially during wintertime when the sun disappears behind the mountains from November until February, and blizzards can cut off the village from the rest of civilisation for days. Bjartur knows that one can never influence the forces of nature, and although most city dwellers visiting the Westfjords in the summer cannot imagine living in a remote place like this, for Bjartur it is just his *way of life*, being his own master on his small boat,[3] a childhood dream that eventually came true. The web of relations and practices through which Bjartur sails today, however, has changed dramatically in the past 30 years, and with it also the discourses and practices that constitute Bjartur as independent coastal fisher.

Not only Laxness's stubborn small-scale farmer, but also our independent small boat fisher reminds us of the struggle for independence that is deeply inscribed in the cultural semantics of the small island state that was settled by Norwegian tax refugees around 850 AD. As far back as Bjartur can remember it has been his childhood dream to make a living on his own boat based on hard but honest work at sea, just as his family and ancestors have done over the generations. Being born just a few years after the first major crisis of the Icelandic fishing industry in the 1960s when the herring fisheries collapsed, Bjartur grew up in a time when small boat fisheries were being rebuilt on the cod- and lumpfish fisheries that were mainly a seasonal, though widespread

---

[2]I will use the forms 'fisher' and 'fishers' throughout the text in place of the traditional 'fisherman' to account for the few women in the industry. For the general role of women in the Icelandic fisheries see Skaptadóttir (1996) and Willson (2016).

[3]In Iceland, coastal fishing vessels are commonly referred to as *small boats* (smábátar). In the following, I will therefore use 'coastal fisheries' and 'small boats' synonymously.

business in local communities. In those days, however, large stern trawlers brought prosperity and wealth to rural coastal areas all around the country. As a consequence of unregulated fisheries and increasing capture capacity, however, fishing pressure grew too high and fish stocks started dwindling by the beginning of the 1980s. As an emergency measure, access to the fishing grounds that used to be a commons of the Icelandic people was closed. Consequently, a quota system, which allotted relative amounts of the Total Allowable Catch (TAC) based on historical catch records was implemented on a trial basis in 1984. As a result, the fish stocks literally became privatised overnight, and it did not take long for vested interests to secure the emergence of a market on which rights to future catches, called Individual Transferable Quotas (ITQs), could be traded just like any other asset from the early 1990s.

Bjartur's family and most of the other members of the community in the Westfjords soon began to fear that this development would be detrimental for the rural areas of the country. A market-based system, the main concern was, could adversely affect the countryside by giving a competitive advantage to big capital holders around the capital region. Nevertheless, during his teenage years Bjartur's community was thriving as two larger longline vessels provided many jobs, and it was still common for locals to run their own small coastal vessel as part of their seasonal full- or part-time occupation. Soon, Bjartur started helping out as deckhand on his parent's small coastal vessel during his school holidays, which remained—as were all coastal vessels below 15 metres in length—widely protected from the regulatory framework of the ITQ system. After graduating from school, Bjartur finally attended navigation school in Reykjavík to becoming a captain himself and later made it to first mate on a large fishing vessel. But even though the pay was extraordinarily good, life on board a huge swimming metal box that included long trips away from home and his girlfriend soon reminded him of his childhood dream of being independent and his own master on a small fishing vessel based in his home community.

Recent developments in the fishing industry made this development even more attractive to him: the emergence of fish auctions in the late 1980s gave independent coastal fishers a much better standing and higher prices per kilo as they have liberated them from the power of the local processors. On the other hand, legal loopholes allowed large-scale investors to buy into the coastal fleet and start operations, ironically leading to a revival of the coastal fisheries. Hence, it was just a matter of time before the government started to implement stricter regulations, and many coastal fishers who had not been fishing much lately on their own vessels feared losing their right to fish as catch quotas were allocated based on historical catch rights with the implementation of the mother ITQ system in 1984. It was therefore about time for Bjartur to quit his job and start to pursue his dream on his parent's old wooden fishing vessel.

With increasing capitalisation and professionalisation of the coastal fleet, many of the newer vessels were able to fish all year round, and above all, the efficiency of state of the art capture technologies increased season by season. As expected, it was just a matter of time before the government adopted new regulations for the coastal fisheries. As a consequence, the regulations started moving successively towards the market-based rationale of the ITQ system, going from a period of a limited numbers of *days at sea* for each vessel to a fully-fledged coastal ITQ system for all vessels under 15 metres. Eventually, in 2004 fishing rights could be fully transferred or leased between small boats, independently from a vessel's homeport and geographical location.

Based on their historical catch records, Bjartur and his family were allotted a relative quota share that literally overnight turned them into members of a privileged class with access to the nation's most valuable resource, being totally free to decide whether they wanted to buy more quota or sell out of the industry. In contrast to others from the community, however, the latter option was out of the question for Bjartur and his family with regard to what they have built up over the years. And besides that, it seemed that fishing on small boats had never been as comfortable before: in contrast to the olden days of 'derby style' fishing, in which small boat owners had to race against their competitors at sea, making them extremely vulnerable to taking risks in bad weather, the limitation of fishing rights had already secured the catch for the season; Bjartur simply needed to decide when to put to sea and harvest it until his quota for the season was used up.

Image 1.1   The 'small boat revival' in Bolungarvík, Westfjords (Photo by AD)

Today, Bjartur and his family consider themselves lucky as they have so far managed to keep their independent business afloat. And in contrast to the hot air created by the financial alchemists in the capital region, the fish remained not only relatively abundant and real in the sea, but also brought in fresh foreign currency in the booming export industry—the main reason why Bjartur's sister decided to move back from the financially devastated capital region to the Westfjords to find a job in the fishing industry after the 2008-financial crisis. In the neighbouring village, the economic success of the coastal fisheries has even lead people to talking about a 'small boat revival' (see Image 1.1), which was also the main reason which attracted my scholarly interest in this otherwise remote and desolated place on earth. But is it really only the stubbornness and hard work of Bjartur and his rural fellows at sea that accounts for their success?

Coastal Fisheries—A Polanyian Countermovement?

Against all the odds of the classical sociological canon in which the 'satanic mills' (Polanyi, 2001: 77) of unfettered markets and industry tend to infiltrate and destroy more harmonious forms of community life,[4] the story of Bjartur shows that small artisanal and labour intensive industries have not only survived the modernisation of the fishing industry by the turn of the century, but also the closure of the marine commons through market-based reforms in the 1980s, as well as the financialisation of the Icelandic economy in the 1990s and the respective financial crisis of 2008. Moreover, instead of being wiped out by more efficient forms of resource allocation and market competition, some have even managed to rebuild their local economies around small boats, bringing jobs and hope to otherwise desolate rural regions. So can the unexpected revival of the Icelandic small boat fisheries be viewed as a sign of rural resilience and liberation—the final episode of an independent people against global capitalism—a victory of David over Goliath? Put differently, has the implementation of new markets and property rights sparked a new Polanyian countermovement that has successfully limited the devastating consequences of neoliberalism in the rural periphery?

The idea of small and labour intensive forms of production as viable alternatives to large-scale industrialism can look back to a long history in both conservative and liberal-green movements, respectively. Already by the turn of the nineteenth century, the idea of local and labour intensive forms of subsistence production was growing in popularity in anti-modern circles and came to global acclaim through Knut Hamsun's (1921) Nobel prize winning novel *Growth of the Soil* in which family farmer Isak overcomes the modern promises of bottomless expansion through daily hard labour in harmony with his land. While Hamsun's archetype of the

---

[4]See especially Hirschmann's (1982) overview of the 'destruction-hypothesis', according to which markets tend to undermine traditional forms community-life. A similar view is put forward by Habermas' (1981) theory of the 'colonialisation of the lifeworld' through the cold and calculating logic of societal sub-systems, in particular bureaucracy and economy.

sturdy small-scale farmer appealed to the zeitgeist of fascist *blood & soil* ideologies of the 1920s and 1930s, it nevertheless saw a late revival in the second half of the twentieth century in which the destructive social and environmental impact of modern agriculture was becoming more apparent. In particular American economist E. F. Schumacher (1973) further popularised the view of small-scale industries as more humane, empowering and sustainable alternatives to the world of world of large-scale industrialism. Today, the idea of small industries as viable alternative to industrial production and neoliberal globalism has yet again seen a revival with regard to public and academic discourse on rural resilience and development.[5] But is are small and local forms of production per definition 'beautiful', as the romanticising title of Schumacher's bestseller promises? Are we really seeing a new social movement led by *homo ruralis*,[6]—the countryside-cousin of Ralf Dahrendorf's (1973) famous *homo sociologicus* who's role and actions are motivated by community-values that reflect a more sustainable, local and decelerated alternative to the socially and environmentally destructive imperatives of industrial modernity?

The observation that small-scale forms of production can co-exist besides large industrial producers is not new as such and was first noticed in the first quarter of the early twentieth century by Russian agronomist Alexander Chayanow's (1986) *The Theory of the Peasant Economy*, according to which pre-industrial forms of production such as the family farm gain a competitive advantage over large-scale producers due to their ability to flexibly adapt and overexploit their workforce. In this sense, the peasant economy caters industrial capitalism with cheap labourers and raw materials. While Chayanow's ideas were rediscovered rather late by Western academics dealing with issues around global poverty in the second half of the twentieth century,[7] it led some scholars to revive the idea of small and labour intensive forms of production as key to rural

---

[5]E.g. Shucksmith and Rønningen (2011) stylise small-scale industries as countermovements to neoliberalism.

[6]In this case, *homo ruralis* represents the ideal type of the community-bound and quasi-subsistent rural dweller.

[7]For an overview of the debate around peasant economy, see Harris (1982).

development. Others, in particular Marxist scholars, however believed that the survival and political protection pre-capitalist forms of production are major obstacle to economic prosperity and growth. From this perspective, the romantic myth of homo ruralis appears as 'neo-populist pipe dream' (Byres, 1979) working against the developing aspects of industrialisation. From this perspective, there can only be one possible explanation that accounts Bjartur's survival: reactionary policies including monetary compensations and laws that maintain a widely popular but inefficient and outdated industry for ensuring the delivery of cheap raw materials and stable resource rent by a more or less impoverished peasantry. A similar aversion against the romantisation and protection of small industries can be expected from liberal economists suspecting the state to waste 'taxpayer's money' in order to artificially maintain an inefficient and wasteful tradition that hampers economic growth in an unattractive work environment.[8]

But although some public money is spent on keeping fisheries-related jobs in some deprived coastal communities, the answer cannot be found merely in state support and tariffs that still protect large parts of European agriculture and fisheries. It therefore might come as a surprise to the reader that Bjartur is financially standing more or less on his own two feet. Moreover, Bjartur and his friends do no longer simply serve the centres of global capitalism with cheap protein. Nor do they seem to follow their rural ancestors by mobilising 'politics against markets' (Esping-Andersen, 1985). Rather it seems as new markets and property rights have transformed coastal fishers into a new class of petit-capitalist entrepreneurs and investors who cater a new, exclusive luxury niche-market for 'sustainable' and 'fresh' fish products. In fact, Bjartur and the other small boat owners have kindled a new entrepreneurial spirit that has *revalued* coastal fisheries as highly valuable and profitable industry that has even attracted interest from the financial institutes in the capital region (more on this later). In other words, the world of Bjartur has little to do with the world of the low-tech rural peasantry of Chaynaow, Hamsun, Schumacher and others.

---

[8]See for instance Hannesson (2004: 134) for this view on rural development.

So can Bjartur's metamorphosis rather be interpreted as a story of liberation, symbolising the metamorphosis from the poor, inefficient and community-bound *homo ruralis* to *homo oeconomicus*, the free and rational fisher who has become truly *independent* from his rural ties? Must we, after all, even thank market ideologues for finally writing the happy ending for Laxness's paragon, Bjartur, the independent and prosperous fisher who lives happily ever after?

It seems that we must admit that sociological and anthropological theories of peasant economies have a hard time explaining Bjartur's transformation from rural peasant to entrepreneur and investor. On the other hand, can it be explained merely by introducing economic man to the fleet?

Fishing in Liberal Capitalism

In contrast to his predecessor coastal fishers for whom fishing was merely a seasonal occupation, Bjartur exemplifies ideal-typically the contemporary, though dwindling species of the independent Icelandic coastal fisher who is able to freely coordinate his fishing activity all-year round with regard to his fishing quota, changing market prices, fishing techniques, weather conditions and fishing seasons. Not to say that Bjartur's struggle for independence is less arduous than that of the peasants in the olden days, but ways of engaging with his world have changed tremendously with the implementation of a market for fishing quotas in 1984 and the growing techno-scientific apparatus attached to it.

Historically speaking, trade and markets as such are nothing new to the coastal settlements of the North Atlantic and we have good reasons to question if the hermetically sealed fishing community has ever existed.[9] Nevertheless, the modernisation of the North Atlantic fisheries

---

[9]Especially the development of individual farmsteads and seasonal fishing stations to more stable settlements and towns in the 19th and twentieth century rather seems to be the result of increasing commercial activities and trade, not their decline (Dobeson, 2019, forthcoming). Moreover, British anthropologist J. A. Barnes (1954), pioneer of social network theory has noticed in his early study of a modernising Norwegian island parish the importance of interpersonal ties beyond the geographical boundaries of the community, which are of particular importance for recruiting seasonal labourers for the fishing industry.

as well as the emergence of industrial capitalism occurred rather late with the advent of motor and freezing technologies that lay the foundation for the introduction the Fordist production model and the emergence of global consumer markets for frozen fish after the Second World War (Apostle et al., 1998: 59–84). The period of economic prosperity that industrial capitalism brought to the coastal communities, however, was only short-lived and ill-fated from the start. While coastal fisheries maintained an important role for coastal communities by buffering the high economic risks of large-scale industrialism, ever-efficient trawlers and bottomless global demand for cheap protein brought local fish stocks below reproduction levels to near collapse in the second half of the century, calling for more political solutions to prevent further environmental and economic damages.

Today, the fisheries of the North Atlantic are largely organised around Garret Hardin's (1968) influential article about *The Tragedy of the Commons*, according to which rationally acting individuals will be deplete a common-pool resource if not regulated and policed by a strong state. While the emergence of this new 'top-down' management regime has restricted access to marine resources by moratoria and later by the allocation of fishing quotas, it has been widely criticised by fishers, local communities and social scientists for being widely anti-democratic and ignoring the local contingencies of fisheries management, thus calling for alternative solutions for empowering homo ruralis with reforms around self-organisation and co-management (Acheson, 2003; Jentoft, 1989; Ostrom, 1990).

Following the neoliberal zeitgeist of the 1970s and 1980s, however, policy makers seem to have been more convinced by economist's call for reintegrating their problem child homo oeconomicus to the fleet. From a neoclassical economists' perspective the problem lies not in the selfishness of economic man as such, but the institutions that govern his behaviour. From the perspective of market-based resource management, Bjartur's predecessors were lacking the right incentives to unfold their full potential as rational utility-maximising individuals. Hence, it doesn't come as a surprise that, from a neoclassical economists's point of view, the most suitable institution for disciplining Bjartur and ensuring the most efficient allocation of fishing quotas is 'the market', which will sort out the most efficient owners in the long run (Hannesson, 1991). Today, resource economists even claim that ITQs bear the wonderful potential to create 'new wealth' when used as collateral that stimulates investments inside- and outside the fishing industry (Arnason, 2008).

Economic sociologists have noted, however, that markets, are not simply the result of a spontaneous coming-together of individual actors, but dependent on a high degree of social organisation, in particular by the state who enforces property rights and sets the terms of trade (Ahrne, Aspers, & Brunsson, 2015). The same can be said about markets for fishing rights, in which the state controls quantities and allocation of rights. Moreover, and unlike most conventional markets, the making of ITQ-markets requires a complex techno-scientific knowledge regime that translates fish and fishers into the abstract language of the neoclassical market model. Key to this process is what Çalişkan and Callon (2010: 2) refer to as *economisation*, i.e. 'the process through which behaviours, organisations, institutions and, more generally, objects are constituted as being "economic"'. Accordingly, it is not important if the neoclassical market model is empirically 'accurate' or 'true' in the ontological sense. Rather, it is merely necessary that the model treats fish and fishers *as if* scientific models about human and aquatic behaviour were true (Holm, 2007). While it is relatively easy to count the number of trees in a forest, however, counting an unknown number of fish in the sea requires a complex techno-scientific alliance of marine biologists, population ecologists, research vessels, mathematical models and algorithms that lay the bedrock for ITQs by providing predictions about the current state and future availability of fish stocks (Holm & Nolde Nielsen, 2007).

While this new techno-scientific knowledge regimes builds the foundation for the organisation of new markets and property rights, it has at the same time fundamentally changed the ways in which Bjartur engages with his environment. Today, he must not only have an eye on his competitors at sea, but also on changing regulations, market prices, weather conditions, political decisions, as well as interest and exchange rates of international currency markets. Put differently, it is no longer enough for Bjartur to be a skilled fisher, as he has to engage in a variety of different territorial arrangements in order to stay afloat. In this sense, we can say that Bjartur's relations with world have surpassed 'good old days' of the locally bound fishing community to a broader globalised web of new markets, property-rights, international financial markets, scienific discourses and technological development.

In the new world of market-based resource management, however, Bjartur's independence is conditional. Not being policed directly by the state, the top-down management regime of modern resource management has given way to a new decentralised and incentive driven governance regime of *liberal*[10] rural capitalism, in which the 'veridiction of the market' (Foucault, 2009: 32) and not the redistributive solidarity of the state decides who stays afloat. In other words, Bjartur can no longer rely on the help from his home community; nor can he count on the states' rural development policy. One could say that Bjartur has become rather lonely in his struggle for rural independence.

While we now have gotten a first glimpse into what has triggered Bjartur's metamorphosis, the mystery of his survival has grown even bigger. So how can small and labour-intensive industries stay afloat in the volatile world of liberal capitalism?

Preparing for Departure

In order to solve the puzzling case of Bjartur's epic quest for independence, I invite the reader to continue her journey to the rural periphery of liberal capitalism. Although this drastic step will require us to discover some potentially unknown waters (both empirically and theoretically), it seems necessary considering the insufficient alternatives at hand. To do so, we thus need to leave our urban comfort-zone behind, travel to the modern rural landscapes and get down to the nitty-gritty of daily economic life in order to gain a deeper understanding of *how* small boats navigate liberal capitalism. Spending some time at multiple sites reaching from fishing communities, harbours, fish auctions and fishing vessels will decentralise our view to a more integrated perspective

---

[10]For a more general account of this new regime of liberal capitalism, see Münch (2012).

that can be summarised in the light of what Desmond (2014) has summarised under the label 'relational ethnography'. Hence, we do not, as compared to traditional ethnographies, intend to limit our fieldwork to observing and comparing clearly defined groups or places, nor do we have to 'penetrate' a social field by 'going native' and giving up the boundaries between researcher and local culture. Instead, we shall place ourselves as a participating outsider who aims at observing and understanding the 'configurations of relations' (ibid.: 554) between multiple human and non-human actors and sites such as fishers, fish, vessels, auction and quota markets, processors and fishing communities that altogether make up the world of modern coastal fisheries. We consequently do not intend to develop a strict analytical framework for comparing the development of different 'winner' and 'loser' communities. Nor do we aim at unveiling some sort of latent objective forces that drive the rural transformation. Rather, we intend to decentralise Bjartur's metamorphosis by exploring how the broader global entanglements of local practices with new markets, property rights and technologies that transform and reconfigure the web of meaningful relations that altogether revalue the meaning of small boats in the new culture of liberal rural capitalism.

Instead of simply conceptualising Bjartur as helpless marionette in a given social structure or conventionalising him as truly 'free' economic agent and market actor, our journey follows the 'practice turn' (Schatzki, Knorr Cetina, & Savigny, 2001) that breaks with dualistic conceptions of agency as proposed by conventional action theories.[11] Hence, if we assume with Schatzki (2001: 53) that 'social orders are arrangements of people and of artefacts, organisms, and things through which they coexist, in which these entities relate and possess identity and meaning' and that these orders are 'instituted within practices' (ibid.: 45), we can shift the focus of our analysis to the contingent situatedness of daily human coping with others and things in which the

---

[11]In contrast to conventional action theories or structuralist accounts, practice theory rejects essentailising notions of 'the subject'. Instead, practice theoretical accounts are interested in studying how subjectivity and social order are constructed and reproduced as discourse in bodily and socio-material practices (also see Reckwitz, 2002; Schatzki, 1996).

social order of liberal capitalism is anchored and materialised. Thus, like fish that can move with the slack of the fishing line, fishers can move in relative freedom within the boundaries of their socio-material entanglements. And like a fish that manages to shake off the hook in its struggle for survival, fishers can either sell out of the industry, or must learn to economise their energy and find alternative ways of coping in order to stay afloat. In the latter case, however, both fish and fishers need to be aware that the hooks will hold on more tightly the more pressure they put on the lines when trying to expand the boundaries of their freedom.

We will travel to a region in which they abound and the consequences of liberal capitalism are displayed as sharply as possible: the rural Westfjords. Off the beaten track of the ring road, which connects the villages and towns around the island, the landscape, with its numerous fjords and rugged coastline, sticks out from the mainland like a giant claw, with its fingers stretching into the Arctic Ocean, while pointing towards the east coast of Greenland, at roughly 280 kilometres' distance. Hence, lying in the outer northwest of the island state in the middle of the North Atlantic, the impression shared by most urban dwellers when arriving in the Westfjords is probably best described as 'remote'. This feeling is usually amplified during the winter months when Atlantic storms and heavy snow falls occasionally shut down both car and air traffic and increase the danger of avalanches from the numerous mountains along the coastline, behind which the sun disappears for months over the winter.

The harshness of the environment has also shaped the local economies, which almost exclusively depend on fisheries: being geologically the oldest and northernmost parts of the country just below the Arctic Circle, the Westfjords are a comparatively 'cold' region due to the delayed and even shorter summers, in which the midnight sun often hides behind the clouds hanging over the mountains and the lack of hot springs, making agriculture even less an option than in the rest of the country. With the implementation of the quota system in the 1980s, however, the region suffered significantly as the bulk of fishing rights were moved to larger companies in the capital region due to better infrastructure, leading to an overall decline in population for the region.[12] Despite this general economic decline and struggle, however the Westfjords region traditionally holds the largest amount of fishing vessels due to its high number of small boats (see Figs. A.1 and A.2 in Appendix). Hence, after the general decline of the trawler industry in the 1990s (see Fig. A.3 in Appendix), coastal fisheries started to provide the main source of labour and income in many rural economies relying on nothing but fisheries, such as the Westfjords. Thus, the socio-economic environment of the Westfjords and its dependency on fisheries and small boats makes the region the most obvious case for studying the consequences of market-based resource management.[13]

The empirical material presented was collected during a total of 16 weeks of ethnographic fieldwork between April 2012 and July 2014. The dataset comprises 31 semi-structured in-depth interviews ranging from one to four hours and a comprehensive collection of field notes, including 62 informal 'on the spot' interviews with a broad range of stakeholders, including quota-owners, skippers, deckhands, representatives of fish auctions, harbour managers, processors, factory workers and other members of the communities. Furthermore, 55 observations from fishing boats, processing plants, harbour facilities and general community life were documented as field notes.[14] Furthermore, statistics, reports, legal texts and websites on fisheries-related issues that complement the ethnographic part of observation. I have revisited the study site between 2016 and 2018 on numerous occasions in order to keep up with the local and regional developments and updated the data accordingly.

---

[12]Since 1998, the population in the whole Westfjords region has decreased steadily, from 8656 in 1998 to 6870 in 2017 (Statistics Iceland, 2017).

[13]The fieldwork conducted for this study was conducted mainly around the municipalities of Bolungarvík and Ísafjörður, where the main fishing activities in the region take place. The municipality of Ísafjörður is the administrative centre and the most populated place in the Westfjords, with a total of 3748 inhabitants (Statistics Iceland, 2013b), including the fishing villages Flateyri, Þingeyri, Hnífsdalur and Suðureyri. Despite its size of only 2624 (Statistics Iceland, 2013a), the town has a fairly urban infrastructure and atmosphere, including a town centre with shops, bakeries, a hotel and cafés, a town hall, guest houses, a library, a hospital, a pharmacy, a swimming bath, a fish monger, two supermarkets, several restaurants, a museum, a small airport and is the only place offering higher education in the region.

[14]Quotes from the transcripts will be marked *M* for manuscript, followed by its number in the data set, which is marked with a roman digit; field notes will be quoted as FN.

Why Study Small Boats?

After all, the critical reader may ask: why should we be bothered about a bunch of small fishing vessels on an island with a population of roughly 350,000. Shouldn't we rather invest our energies in studying the transformation of large industries in the urban centres of global capitalism?

Social scientists dedicated to understanding the institutions and dynamics of contemporary capitalism have spent much time studying issues around industry and labour (Esping-Andersen, 1990; Hall & Soskice, 2001; Hollingsworth & Boyer, 1997), and more recently modern finance (Davis, 2009; Knorr Cetina & Preda, 2005) and aesthetic markets in different empirical fields such as art (Velthuis, 2005) and fashion (Aspers, 2010). Despite their broad empirical variation, all these industries have in common that they are situated in the urban centres of modern capitalism. This 'urban bias', however, tends to oversee the importance of the rural hinterlands, in particular natural resource-based economies as important generator of consumer goods and wealth for the urban centres as such (Cronon, 1991).

Furthermore, although scholars in the field of rural studies have paid considerable attention to the consequences of neoliberal restructuring (Shucksmith & Brown, 2016) and the emergence of a new global 'corporate food regime' (Busch, 2010; McMichael, 2005; Pechlaner & Otero, 2010), rather little is known about its link to local practices of production, valuation and exchange in rural economies (for an exception, see Çalişkan, 2010). What is furthermore ignored in this transformation of global food production is the role of traditionally labour intensive and small industries, which tend to be seen merely as pre-modern or subsistent forms of production with little importance to global capitalism as such. Hence, small industries tend to be recognised in their role for poverty alleviation and rural development (Shucksmith & Rønningen, 2011). While these accounts rightly claim that small industries are 'too big to ignore' (Chuenpagdee, 2011), however, they likewise seem to ignore the important role of small industries in the broader context of neoliberal reform, rural restructuring and 'green consumerism' (Boström &

Klintman, 2011). In the light of these shortcomings, this book aims at showing that small industries, in particular coastal fisheries are in particular interesting in the study contemporary capitalism they show how notions of 'small', 'local' and 'sustainable' themselves become important vehicles of value generation and profit-making. Moreover, small industries are ideal study sites that mirror the dynamics, changing geographies of capital and cultural conflicts between the rural peripheries and the cosmopolitan centres in contemporary global capitalism.

All in all, and in line with the thriving interdisciplinary field of valuation studies (Antal, Hutter, & Stark, 2015; Beckert & Aspers, 2011; Lamont, 2012; Stark, 2009) studying the transformation of small and labour intensive industries in the wake of neoliberal reform opens up wide range of issues of interest to scholars and graduate students in the field of science and technology studies and economic sociology related to issues of economisation and marketisation in society (Beckert, 2009; Çalişkan & Callon, 2009, 2010; Schimank, 2014; Schimank & Volkman, 2012). Finally, the case of the Icelandic coastal fisheries will give scholars, students and policy makers interested in the role of market-based solutions to issues around sustainability and rural development a deeper understanding of the dynamics and consequences of neoliberal reform in the countryside.

Itinerary

In order to circumvent any disappointment for the time invested in reading this book, I will start this summary with a disclaimer. The reader should not expect a book defending the romanticised ideal of the organic fishing community against a bunch of greedy capitalists and bankers. The reader will soon learn that identifying the villains is not that simple. Nor does the author intend to promote artisanal fishing techniques and small-scale fisheries as especially 'sustainable' or 'environmentally friendly'—judgement on these important issues is surely beyond the modest capabilities of the sociologist. In the end, the reader may judge for herself to what extent the distinction of 'small' vs 'large' still makes sense and how 'beautiful'[15] today's small coastal vessels really are.

The reader can however expect to dig deeper into the story of Bjartur's metamorphosis by providing an encompassing account of the social dynamics, practices and tensions that characterise and challenge the boundaries of today's rural economies in the wake of a new culture of liberal capitalism. In order to do so, more ethnographic material from Bjartur and his fellow Icelandic coastal fishers is meshed into the storyline and discussed in relation to accounts and theories dealing with issues such as economisation, marketisation, natural resources, technology and valuation. The structure of the book is as follows.

Before we return to the main plot, Chapter 2: *Fishing in Market-Based Resource Management* discusses the rise of market-based resource management as dominating paradigm in contemporary resource economics. We will critically assess to what extent existing social scientific research can help us understanding the transformation of contemporary fisheries in the wake of market-based reforms. We ask: What general insights have social scientists generated about the role of markets and property rights in fisheries? And what conceptual blind spots can be identified?

Chapter 3: *Economising Rural Independence* finally brings us back to the main plot and investigates how new markets and property rights transform rural production networks. We ask: what social and cultural role have coastal fisheries played historically? And how has it changed with the privatisation of access rights and the organisation of new markets?

Any account of the fishing industry would remain incomplete without taking into account the practice that lies at the heart of the value chain: fishing. For this reason, Chapter 4: *The Practice of Fishing* takes the reader onto a day's adventure on a coastal fishing vessel and provides a first-hand experience of the practice of fishing itself. We will reflect on questions such as: how can we describe the relation between the fishers and their environment? How do fishers coordinate their work at sea?

---

[15]The idea that ‚small is beautiful' was promoted by the economist J. F. Schumacher (1973), who contrasts larg-scale economies and technologies with an allegedly more humane, empowering and sustainable world of small-scale economies.

Chapter 5: *Enframing the Sea* investigates the role of digital technology in modern fishing and theorises how it has changed the relation between fish and fisher, and more broadly our conception of the natural environment as resource that can be utilised, extracted and controlled.

We will soon learn that modern market-based fishing is no longer an isolated activity that is merely based at sea. For this reason, Chapter 6: *When the Fish Ignore the Market* investigates the tensions between the economy and an ever-changing and unpredictable more-than-human environment. We ask: what practical problems do coastal fishers face resulting from the organisation of new markets and property rights? And how do fishers cope with this increasing complexity in modern fishing?

While the practices of daily coping show how fishers, quota owners and processors deploy flexible strategies that allow them to stay afloat in this rapidly changing world, we will learn in Chapter 7: *Fishing for Quality* why 'quality' has become the key issue for understanding the survival of the Icelandic coastal fisheries. For this reason, we will first investigate the general role of quality in liberal capitalism. We ask: Is quality a natural feature of the raw material? And what does all of this imply with regard to processes of economisation and marketisation?

Although quality upgrading is the key to revaluing this traditional rural industry, it has yet again opened up new uncertainties in the production network. In Chapter 8: *The Fishery Panopticon* we will learn about the importance of digital information technologies for the new quality-regime. We ask: how do buyers in electronic auction markets ensure that the quality of the raw materials they are going to purchase is up to their standards? And what does the digitalisation of the supply chains imply for the practice of fishing?

Chapter 9: *A New Liberal Culture of Rural Capitalism* concludes and asks to what extent our stubborn anti-hero of the rural coastal fisher has managed to defend his independence—or if he is doomed to fail, like his literary cousin.

Last but not least, the *Appendix* will give the unfamiliar reader the chance to get acquainted with the basic capture methods used in the Icelandic coastal fisheries, which will play an important role in the struggle for rural independence.

Back to Business…

This morning, fish fever has taken full control of Bjartur. First, he will take a look outside the window in his living room, which allows an encompassing view over the fjord. The moonlight shines brightly on the snow, making the shapes of the surrounding mountains and some scattered clouds visible to his sleepy eyes. But the mountain to the east side of the fjord seems clear and no clouds are covering its peak, which indicates better weather, as experience has taught him. He opens the window and carefully listens to the gently breeze moving through the village's streets, leaving unassigned clatter and occasional howls. Although the worst part of the storm seems to have moved away from his home, he knows that the sheltered fjords should not fool him, since weather conditions at sea can vary tremendously within miles. And even if the fjord looks calm, it can literally be like sailing straight into a wall of strong winds and giant swells at its mouth. For this reason, he switches on his notebook and checks the website of the Icelandic Marine Administration, where he will get the latest information on wave height and intervals per second, two factors that Bjartur evaluates against each other. At the same time, a look at the vessel tracking website marinetraffic.com lets him see whether any of his colleagues around the country have already put to sea—and indeed he can see that two of the newer bigger longliners owned by a larger processing company from a neighbouring village have already left port. Although Bjartur does not trust these vessels, as they are infamous for being risk takers—some rumours say they are just 'crazy' people in this village, others say it is because of the pressure from the owners—Bjartur knows that he cannot compare his boat to these considerably larger 'small boats' weighing up to 15 tonnes. The wind forecast, however, looks good for the day and the waves seem to have flattened out so that Bjartur decides to call his deckhand and tell him to pick up the fishing lines from the baiting house and meet him at the docks in about thirty minutes…

References

Acheson, J. M. (2003). *Capturing the Commons: Devising Institutions to Manage the Main Lobster Industry*. Lebanon, NH: University Press of New England.

Ahrne, G., Aspers, P., & Brunsson, N. (2015). The Organization of Markets. *Organization Studies, 36*(1), 7–27.

Antal, A. B., Hutter, M., & Stark, D. (Eds.). (2015). *Moments of Valuation*. Oxford: Oxford University Press.

Apostle, R., Barret, G., Holm, P., Jentoft, S., Mazany, L., McCay, B., & Mikaelsen, K. (1998). *Community, State and Market on the North Atlantic Rim: Challenges to Modernity in the Fisheries*. Toronto: University of Toronto Press.

Arnason, R. (2008). Iceland's ITQ System Creates New Wealth. *The Electronic Journal of Sustainable Development, 1*(2), 35–41.

Aspers, P. (2010). *Orderly Fashion: A Sociology of Markets*. Princeton: Princetion University Press.

Barnes, J. A. (1954). Class and Commitees in a Norwegian Island Parish. *Human Relations, 7*(1), 39–58.

Beckert, J. (2009). Wirtschaftssoziologie als Gesellschaftsstheorie. *Zeitschrift für Soziologie, 38*(3), 182–197.

Beckert, J., & Aspers, P. (Eds.). (2011). *The Worth of Goods: Valuation & Pricing in the Economy*. Oxford: Oxford University Press.

Boström, M., & Klintman, M. (2011). *Eco-Standards, Product Labelling and Green Consumerism*. Basingstoke: Palgrave Macmillan.

Busch, L. (2010). Can Fairy Tales Come True? The Surprising Story of Neoliberalism and World Agriculture. *Sociologia Ruralis, 50*(4), 331–351.

Byres, T. J. (1979). Of Neo-Populist Pipe-Dreams: Daedalus in the Third World and the Myth of Urban Bias. *The Journal of Peasant Studies, 6*(2), 210–244.

Çalışkan, K. (2010). *Market Threads: How Cotton Farmers and Traders Create a Global Market*. Princeton: Princeton University Press.

Çalışkan, K., & Callon, M. (2009). Economization, Part 1: Shifting Attention from the Economy Towards Processes of Economization. *Economy and Society, 38*(3), 369–398.

Çalışkan, K., & Callon, M. (2010). Economization, Part 2: A Research Programme for the Study of Markets. *Economy and Society, 39*(1), 1–32.

Chayanov, A. V. (1986). *The Theory of Peasant Economy*. Madison: The University of Wisconsin Press.

Chuenpagdee, R. (2011). Too Big to Ignore: Global Research Network for the Future of Small-Scale Fisheries. In R. Chuenpagdee (Ed.), *World Small-Scale Fisheries: Contemporary Visions* (pp. 383–394). Delft: Eburon.

Cronon, W. (1991). *Nature's Metropolis: Chicago and the Great West*. New York: W. W. Norton.

Dahrendorf, R. (1973). *Homo Sociologicus. Ein Versuch zur Geschichte, Bedeutung und Kritik der Kategorie der sozialen Rolle*. Opladen: Westdeutscher Verlag.

Davis, G. F. (2009). *Managed by the Markets: How Finance Reshaped America*. Oxford: Oxford University Press.

Desmond, M. (2014). Relational Ethnography. *Theory and Society, 43*, 547–579.

Dobeson, A. (2019, forthcoming). Das Fischerdorf im liberalen Kapitalismus: sozialräumliche Öffnungs- und Schließungsprozesse in der nordatlantischen Peripherie. In A. Steinführer, L. Laschewski, T. Mölders, & R. Siebert (Eds.), *Das Dorf. Soziale Prozesse und räumliche Arrangements*. Berlin: LIT.

Esping-Andersen, G. (1985). *Politics Against Markets: The Social Democratic Road to Power*. Princeton: Princeton University Press.

Esping-Andersen, G. (1990). *Three Worlds of Welfare Capitalism*. Princeton: Princeton University Press.

Foucault, M. (2009). *The Birth of Biopolitics: Lectures at the Collège de France 1978–1979*. Basingstoke: Palgrave Macmillan.

Habermas, J. (1981). *Theorie des kommunikativen Handelns. Band 2: Zur Kritik der funktionalistischen Vernunft*. Frankfurt am Main: Suhrkamp.

Hall, P. A., & Soskice, D. (2001). *Varieties of Capitalism: The Institutional Foundations of Comparative Advantage*. Oxford: Oxford University Press.

Hamsun, K. (1921). *Growth of the Soil*. New York: A. A. Kopf.

Hannesson, R. (1991). From Common Fish to Rights Based Fishing: Fisheries Management and the Evolution of Exclusive Rights to Fish. *European Economic Review, 35*, 397–407.

Hannesson, R. (2004). *The Privatization of the Oceans*. Massachusetts: MIT Press.

Hardin, G. (1968). The Tragedy of the Commons. *Science, 162*, 1243–1248.

Harris, J. (Ed.). (1982). *Rural Development: Theories of Peasant Economy and Agrarian Change*. London: Hutchnis University Library.

Hirschman, A. O. (1982). Rival Interpretations of Market Society: Civilizing, Destructive, or Feeble? *Journal of Economic Literature, 20,* 1463–1484.

Hollingsworth, R. J., & Boyer, R. (Eds.). (1997). *Contemporary Capitalism: The Embeddedness of Institutions.* Cambridge: Cambridge University Press.

Holm, P. (2007). Which Way Is Up on Callon? In F. Muniesa, L. Siu, & D. A. MacKenzie (Eds.), *Do Economists Make Markets?* (pp. 225–243). Princeton: Princeton Univeristy Press.

Holm, P., & Nolde Nielsen, K. (2007). Framing Fish, Making Markets: The Construction of Individual Transferable Quotas. In Y. Millo, M. Callon, & Fabian Muniesa (Eds.), *Market Devices* (Vol. 55, pp. 173–195). Malden: Blackwell.

Jentoft, S. (1989). Fisheries Co-management. *Marine Policy, 13*(2), 137–154.

Knorr Cetina, K., & Preda, A. (Eds.). (2005). *The Sociology of Financial Markets.* Oxford: Oxford University Press.

Lamont, M. (2012). Toward a Comparative Sociology of Valuation and Evaluation. *Annual Review of Sociology, 38,* 201–221.

McMichael, P. (2005). Global Development and the Corporate Food Regime. In P. M. Frederick & H. Buttel (Eds.), *New Directions in the Sociology of Global Development* (Vol. 11, pp. 269–303). Bingley: Emerald.

Münch, R. (2012). *Inclusion and Exclusion in the Liberal Competition State: The Cult of the Individual.* London and New York: Routledge.

Ostrom, E. (1990). *Governing the Commons: The Evolution of Institutions for Collective Action.* Cambridge: Cambridge University Press.

Pechlaner, G., & Otero, G. (2010). The Neoliberal Food Regime: Neoregulation and the New Division of Labor in North America. *Rural Sociology, 75*(2), 179–208.

Polanyi, K. (2001). *The Great Transformation: The Political and Economic Origins of Our Time.* Boston: Beacon.

Reckwitz, A. (2002). Toward a Theory of Social Practices: A Development in Culturalist Theorizing. *European Journal of Social Theory, 5*(2), 243–263.

Schatzki, T. R. (1996). *Social Practices: A Wittgensteinian Approach to Human Activity and the Social.* Cambridge: Cambridge University Press.

Schatzki, T. R. (2001). Practice Mind-ed Orders. In T. R. Schatzki, K. Knorr Cetina, & E. V. Savigny (Eds.), *The Practice Turn in Contemporary Theory* (pp. 42–55). London and New York: Routledge.

Schatzki, T. R., Knorr Cetina, K., & Savigny, E. V. (2001). *The Practice Turn in Contemporary Theory.* London and New York: Routledge.

Schimank, U. (2014). Modernity as Functionally Differentiated Capitalist Society: A General Theoretical Model. *European Journal of Social Theory, 18*(4), 413–430.

Schimank, U., & Volkman, U. (Eds.). (2012). *The Marketization of Society: Economizing the Non-Economic.* Bremen: University of Bremen.

Schumacher, E. F. (1973). *Small Is Beautiful: A Study of Economics as If People Mattered.* London: Blond & Briggs.

Shucksmith, M., & Brown, D. L. (2016). Framing Rural Studies in the Global North. In M. Shucksmith & D. L. Brown (Eds.), *Routledge International Handbook of Rural Studies* (pp. 1–26). London and New York: Routledge.

Shucksmith, M., & Rønningen, K. (2011). The Uplands After Neoliberalism?—The Role of the Small Farm in Rural Sustainability. *Journal of Rural Studies, 27,* 275–287.

Skaptadóttir, U. D. (1996). Housework and Wage Work: Gender in Icelandic Fishing Communities. In G. Pálsson & E. P. Durrenberger (Eds.), *Images of Contemporary Iceland. Everyday Lives and Global Contexts* (pp. 87–105). Iowa: University of Iowa Press.

Stark, D. (2009). *The Sense of Dissonance: Accounts of Worth in Economic Life.* Princeton: Princeton University Press.

Statistics Iceland. (2013a). Population by Locality, Age and Sex, 1 January 2011–2013. Ísafjörður.

Statistics Iceland. (2013b). Population by Municipalities, Sex and Age, 1 January 1998–2013—Current Municipalities. Ísafjarðarbær.

Statistics Iceland. (2017). Population by Municipalities, Sex and Age, 1 January 1998–2017—Current Municipalities. Westfjords.

Velthuis, O. (2005). *Talking Prices: Symbolic Meanings of Prices on the Market for Contemporary Art.* Princeton: Princeton University Press.

Willson, M. (2016). *Seawomen of Iceland: Survival on the Edge.* Seattle and London: University of Washington Press.

2

Fishing in Market-Based Resource Management

Before we set sail on our journey to investigate Bjartur's metamorphosis from community-bound rural fisher to independent entrepreneur and investor, it is worth considering the broader social and political context in which the fishing industry has transformed and changed. Hence, especially in the 1970s, the role of the state in organising society has changed dramatically with the neoliberal revolutions of the Reagan and Thatcher administrations of the 1980s a (Harvey, 2005). Following the ideas of influential economists such as Friedrich Hayek and Milton Friedman, this policy shift commonly assumed that the market mechanism is the most efficient means for the distribution and allocation of societal wealth, with the state being reduced to the role of a nightwatchman who oversees and enforces the laws of competition and exchange. Consequently, formerly state run institutions such as transport, education and health care have been fully or de facto privatised in many industrial nations, with even archetypical welfare states such as Sweden following suit the call for privatisation and competition in the public sector (Svallfors & Tyllström, 2018). In the wake of this general 'state phobia' (Foucault, 2008: 75), also the adherence of resource economists and policy makers to market-based environmentalism grew and became

© The Author(s) 2019, corrected publication 2020
A. Dobeson, *Revaluing Coastal Fisheries*,
https://doi.org/10.1007/978-3-030-05087-0_2

the major paradigm to solve a wide-ranging issues of environmental problems caused by pollution and depletion of common pool resources such as air, water and fish (Keohane & Olmstead, 2007).

Market-based approaches to modern resource management commonly assume that unregulated access to resources will inevitably lead to a *Tragedy of the Commons* (Hardin, 1968), as there is no economic incentive for rationally acting individuals to stop putting pressure on a common good unless it becomes either devastated or limited by a regulating leviathan. Key to this reasoning is the conceptualisation of man as *homo oeconomicus*, namely a rational, utility-maximising agent who considers no utility but his own. In other words, economic actors have no other motivation than maximising profits through increasing pressure on a given resource. In order to discipline economic man, resource economists believe that the market mechanism is superior to top-down state control by ensuring that resources are utilised only by the most efficient users. Consequently, the privatisation of the commons and the establishment of property rights is the inevitable precondition for ensuring the sustainable and efficient allocation of natural resources (Hannesson, 2004).

In contemporary fisheries management, market-based solutions to environmental problems have been especially pronounced and widely promoted as best practice solution to the problem of overfishing. Accordingly, proponents of property-rights based approaches see a system of *Individual Transferable Quotas* (ITQs)[1]—freely exchangeable fishing rights that can be traded like any other commodity independently of mother vessel and owner—as the most sustainable and efficient way of utilising a nation's marine resources (Arnason & Runolfsson, 2008; Hannesson, 2004). Today, markets for ITQs build the backbone of many fisheries economies, in particular Iceland (Eythórsson, 1996, 2000) and New Zealand (Hersoug, 2002), which both pioneered rights-based fisheries management reform in its most radical form, while access to fishing grounds in other major

---

[1]A more detailed account of the rationale behind TACs and ITQs will be given in subsequent sections.

European fishery nations such as Norway (Hersoug, 2005), Denmark (Høst, 2015) and the UK (Cardwell, 2015) have been de facto privatised while the EU's Common Fisheries Policy is pushing in a similar direction (European Commission, 2009).

The transformation of the fisheries has gathered much attention from social scientists highlighting different aspects of the changing relation between coastal communities, markets and the state. The purpose of this chapter is therefore to briefly sketch the most prominent contributions to critically assess their explanatory value with regard to the current transformation of rural communities in the wake of market-based reforms. Based on the reviewed literature we will see that the research on modern fisheries follows three main threads found in contemporary economic sociology: *networks, institutions and* socio-technical *discourse*.

Fishing in Social Relations

More than thirty years before Marc Granovetter's (1985) seminal article on social embeddedness was published in the *American Journal of Sociology*, anthropologist J. A. Barnes (1954: 42–43) pondered the relations between what he referred to as the fluctuating, functional and highly competitive *industrial system* of the herring fisheries and the conservative and hierarchical *territorial system* of the community during his field study of an Norwegian island parish. In more contemporary language, Barnes was interested in the social coordination between *markets* and *hierarchies* (Williamson, 1975). Based on his field observations, however, Barnes discovered a unit of social organisation that seemed to be not only different from the aforementioned, but also the missing link that allowed coordination between these seemingly hostile worlds:

> The third social field has no units or boundaries; it has no coordinating organization. It is made up of the ties of friendship and acquaintance which everyone growing up in Bremnes society partly inherits and largely

builds up for himself (...) It is convenient to talk of a social field of this kind as a network. (Barnes, 1954: 43, italics in original)

Although these *ties*, as Barnes (1954: 43) notes, are usually between 'persons who accord approximately equal status to one another', persons with different social status positions, such as fisher and employer, may occur. Barnes (1954) concluded that hierarchies in his Norwegian parish remain rather flat and there were no clear signs of a class in the Marxian sense.[2] It is important to note, however, that Barnes' conception of a network is, in contrast to a two-dimensional 'spider's web', conceptualised with multiple dimensions. Furthermore, the size and quality of the 'mesh', that is, 'the distance around a hole in the network', differs in various types of societies, being rather small in simple and rural, and rather large in modern societies (Barnes, 1954: 44). As Bremnes is described as an 'intermediate society', it has no clearly developed class structure as the network consisted mostly of people that considered each other approximate equals. Class, in this sense, is a merely subjective category. With proceeding industrialisation, however (Barnes, 1954: 57) concludes that 'some form of class society develops as industrialisation proceeds'. This was also evident for Bremnes, where 'the mesh of the network' was already growing larger at the time of Barnes' observations (Barnes, 1954: 5).

More than 30 years later, James M. Acheson (1988) later criticised the Tragedy of the Commons narrative on similar grounds in his seminal life work on the Maine lobster fisheries, which made an important general contribution to contesting the idea of economic man as isolated rational utility maximising agent. His basic argument is rooted in anthropological considerations about the nature of fishing as a socially coordinated activity. In an earlier contribution, Acheson (1981: 276) states that fishing

---

[2]It needs to be pointed out, however, that some Norwegian fishing communities, which were controlled by plant owners resembled more a class society than others, in which fishers were more or less financially independent land owners (I am grateful to Jahn Petter Johnsen for this comment).

takes place in a very heterogeneous and uncertain environment. This uncertainty stems not only from the physical environment, but also from the social environment in which fishing takes place. (ibid.)

Once discovered, an oil stock is fairly controllable and can be exploited relatively independent of the weather conditions. In contrast, weather conditions at sea can change rapidly and pressure skippers to stay ashore or return to their homeports. Furthermore, fish stocks move around, some of them migrate considerable distances, and even if modern fishing technologies have facilitated harvesting, the needle in the haystack has to be discovered first. It is always uncertain where the fish are, and it is always a risk for a skipper to choose a destination for fishing. If his decision proves wrong, he loses time, which represents an economic loss that gets bigger the more competitors are chasing the valuable resource. So how are stability and order possible in this highly volatile and unpredictable environment? Acheson (1981: 307) concludes that while

one cannot control the weather and fish, one can use social ties to organize an effective crew, obtain information on concentration of fish and have privileged access to them, and be assured of a secure market for the catch. (ibid.)

In line with Barnes, this view challenges the economic and folk view of the individualistic outlaw-fisher who fights nature while being in severe competition with his fellows at sea. Rather, Acheson (1977, 1988, 1998) has demonstrated in decades of comprehensive fieldwork and numerous academic contributions how lobster fishers organise themselves as 'harbour gangs' to solve the distributional conflict at sea. Hence, fishing rights are distributed by the gangs, which are tied to specific—though not uncontested—territories and fishing communities and even enforce self-imposed rules on lobster trap limits that have successfully regulated pressure on and secured reproduction of the stocks for decades. Having the right to fish, however, is only half the battle. Due to its informal nature, success 'in the Maine lobster fishing industry depends on ties with fellow lobstermen, the ability to negotiate with lobster dealers, and the sharing of certain skills'

(Acheson, 1988: 1). These ties are densely knit to their homeport communities, which have their own identity, value and kinship systems that can make it fairly difficult for newcomers to enter the fisheries. The same holds true of market relations between producers and fishers, for whom mutual ties are of critical importance for lowering risk and uncertainty in the highly volatile industry. Accordingly, producers 'do not act like economic optimisers, who buy from the lowest-priced source and sell to the highest bidder' as the industry is rather 'characterised by many small enterprises – fishers, dealers, wholesalers, truckers – who have long-standing ties to one another' (Acheson, 1988: 144).

Hence, the lobster industry cannot be described in terms of the neoclassical market model, either, in which prices are the main information provided and contractors only meet for short-term exchanges, nor as a *hierarchy*, in which supply of resources is exchanged between vertically integrated units of production based on long-term relations, as famously distinguished by Williamson (1975). Following Macneil (1978), Acheson states that the industry can rather be described in terms of a *relational market*, in which 'personal relationships strongly influence exchange' (Acheson, 1988: 144). Within this economic environment, identities of fishing communities serve as markers for these territories that go together with the mutual ascription of idiosyncrasies by the members of each community (Acheson, 1988: 27). Moreover, within the community, status hierarchies and kinship determine influence and social position of fishers (Acheson, 1988: 52–57).

In a similar vein, Miller and Van Maanen (1979, 1982) have stressed the heterogeneity of fisher's identities beyond formal classifications of vessel size, fishing gear and target species. In contrast to a *traditional fisher*, who is often 'born' and socialised into his profession and draws clear boundaries between his work at sea and his private life, a *non-traditional* fisher, often referred to as a 'hippie' or 'freak' by the former and typically a younger White Anglo-Saxon Protestant middle class male, is typically as 'prone to wear a rock 'n' roll T-shirt while fishing as he is to wear a fishing knife on his belt while on a date' (Miller & Van Maanen, 1982: 36). Hence, working clothes, fishing gear, storytelling, cars, drinking patterns and even boat names serve as important markers of fishers' identities (Miller & Van Maanen, 1982: 35–38). Especially in multi-ethnic countries such as the United States, status differences between fishers are typically demarcated around ethnic boundaries. Hence, fishers in Gloucester, Massachusetts, who are mainly of Sicilian descent, but consider themselves Americans, distinguish between 'guineas' (themselves) and 'greasers', who represent newly migrated fishers who maintain strong ties to their country of origin and are seen as 'greedy, anti- or at least un-conservation minded, and, in general, uncivilised' (Miller & Van Maanen, 1979: 380).

It has become clear that fishers' and industry's identities are meshed in the communal network of social relations and form important markers for social coordination in fishing communities. The contributions of Barnes, Acheson and Miller and Van Maanen have pointed out the meaning of social relations and identity as backbones of social order in modern fisheries. But how have forms of social coordination changed in fishing communities in which state-organised resource management policies, such as property rights, have been implemented?

Fishing in Institutions

Another branch of sociological research on fisheries emphasises the broader historical context in which modern fishing has been developed and institutionalised. In this light, a number of scholars have studied the consequences of modernisation of the North Atlantic fisheries by putting emphasis on the changing relations between fishing communities, nation-states and market structures. At the core of this transformation lies a shift from the community level towards government control and market rationality at the organisational level. In line with Polanyi (2001), the authors conceptualise modernisation as a *disembedding*, that is, as 'a process through which the economic activities, rationalities, and social relations are "lifted out" from the local context of interaction'. During the process of modernisation, the main argument runs, the rational norms and values of the market sphere have created new institutions that have disembedded the coastal communities from their local environment while becoming the dominant ordering principle in a capitalist society, consequently 'undermining traditional forms of resource management, and eroding the social and cultural bonds on which communities are based' (Apostle et al., 1998: 238–239).

The breakthrough of large-scale industrial capital in the Maritimes, however, did not follow a structural 'logic' of modernisation in which communities necessarily lose their integrative function in favour of the calculative logic of large-scale industrial capital and rationalist production regimes. Rather, coastal communities along the Northern European coastline—such as Iceland and Norway—can be seen as strong empirical cases in which the socio-political and historical context of modernisation was mitigated by their specific socio-political and cultural settings. The cases of Norway shows how the cultural and political context of the industrialisation process has fostered the establishment of a petty capitalist class of fishers that shaped the institutional setting in order to prevent large-scale capitalists for taking over the industrial system during the interwar period (Apostle et al., 1998: 36–58).

Sverisson's (2002) analysis of the Icelandic fisheries points in a similar direction, showing how the Icelandic coastal fisheries coped with modernisation through gradual technological adaption to a specific socio-economic context:

> In Iceland, the technological complex that evolved from the peasant fisheries of the nineteenth century proved particularly resilient. Incremental evolution of the old fishing techniques actually continued to a point where little remained of the original technology. More was retained of the social relations in which that technology was embedded. The small-scale fishermen improved their boats and gear continuously, eventually achieving a comprehensive metamorphosis of vessels and techniques. This happened within a framework defined by strong social and political continuities. (2002: 249)

Thus, the history of the Icelandic coastal fisheries shows that rural fishers rather improve their current gear gradually than replace it radically by more efficient ones. Accordingly, decisions take place within historically contingent trajectories that influence further developments in these industries.

In recent years, however, new forms of market-based management have set off a second wave of modernisation in the fisheries. In the 1970s and 1980s many fisheries across the North Atlantic were brought on the brink of collapse due to systematic overfishing. In reaction to this crisis, many fishery nations reconsidered they ways their fisheries were managed in the light of common property theory, triggering a far reaching transformation from open access to closure of the fishing grounds. Accordingly, access to fishing was either severely limited or even closed by moratoria. While in some fishery nations such as Iceland (Eythórsson, 1996) or New Zealand (Hersoug, 2002) access to fishing grounds was literally privatised over night, institutional reform in other countries such as Norway was confronted with major resistance from the fishing industry. This, however, did not hinder the successive de facto privatisation access rights within a *'a system of individual transferable access rights'* (Hersoug, Holm, & Rånes, 2000: 327, emphasis in original). In contrast to fully transferable ITQs, these so-called Individual Vessel Quotas (IVQs) are constrained to regions and thus cannot be freely traded on a quota market. Thus, the manifestation of property rights in the Norwegian fisheries appears to be the outcome of an historical process, in which actors have been 'locked in' to the historically contingent path in which cognitive, normative and regulative structures of modern resource management lead to gradual institutional change, eventually resulting in a more or less fully-fledged market system of transferable access rights despite strong opposition of key actors in the sector (Hersoug, 2005; Hersoug et al., 2000).

The revitalisation of the notion of 'institutional embeddedness' (Hollingsworth & Boyer, 1997), however, happens at the cost of more general theoretical considerations about the broader role of property rights–based resource management regimes and their anchors in the social contexts in which they are constructed. Besides the fact that coastal communities have been 'disembedded', implying an institutional shift from the community level towards institutions resulting from interaction in markets and the state, it remains unclear how coordination between these three pillars is actually taking place. This has surely something to do with the fact that the term 'market' simply presupposes the neoclassical market model, which consequently is opposed to notions such as 'state' and 'organisation' and often is falsely equated with the ideology of 'capitalism'. Moreover, these conceptual problems point at a more general problem with social scientific accounts of economic phenomena. Thus, the conceptual problems aligned with the embeddedness approach, be it network embeddedness or institutional embeddedness, reflect some general problems with one of the most prominent concepts in economic sociology. As Aspers (2011: 76–77) remarks, it seems as if economic sociological accounts have a tendency of simply adding 'flesh and blood' to economic man, rather than coming up with a theoretically sound alternative to this taken-for-granted dichotomy. In this light, it is no wonder that alternative approaches to fisheries management such as the co-management school (for example, Jentoft, 1989; Hersoug, 1997), which has highlighted democratic decision-making as key to legitimate fisheries management, have yet to developed a promising alternative to the reductionist rationale of contemporary resource economics. We will therefore turn to more integrative perspectives that decentralise economic action by taking into account the broader environment of the economy.

Fishing in Discourse

Although institutionalist accounts have provided important historical studies on the modernisation of the fishing industry and the rise of market-based solutions, scholars associated with the field of Science and Technology Studies have criticised the fact that also multidimensional institutionalist frameworks have not overcome the problematic division between the 'hard' world of rationality, which includes science, technology and market forces, and the 'soft' world of values and norms as the basis for the legitimacy claims of institutions (Holm, 2001, 2007). In a similar fashion, Johnsen (2004) has criticised institutionalist frameworks for explaining increasing investments by fishers in larger vessels despite policy goals of reducing capture capacity as a 'tragedy of soft choices' (Standal & Aarset, 2002), of individual adaptations and adjustments made under the prevailing institutional setting.

Hence, Johnsen (2004: 483) concludes that despite 'being critical to attempts to use the Tragedy of the Commons as the basis for fisheries management, their own understanding of actors ends up being very similar to that of Hardin and his followers' (Johnsen, 2004: 483). Hence, 'more integrated perspectives taking into account the dynamics of technology development, knowledge, policy, and economy are needed' (Johnsen, 2004: 483) in order to decentralise the fishers as the core actors of institutional change. The next two sections will present some perspectives that attempt to decentralise the actor-centric view of both fishers and institutions by taking into account the role of techno-scientific discourses of production.

Fishing in Regimes of Production

An early branch of research on the role of property rights has shed light on the question of how knowledge in different systems of production is reproduced and represented through social practices. Based on these assumptions, Pálsson (1991) has shown how categories such as fish, production and gender are represented as reflected in different 'folk models' of production. In most cultures, for instance, fish serves as a highly ambiguous category loaded with symbolic value and in many fishery economies—not only modern Western fisheries—the work of women is considered as secondary and 'unproductive' in contrast to the 'productive' physical activity of fishing itself, which is widely dominated by men (also see Skaptadóttir, 1996).[3]

Rejecting the classification of systems of production based on technology Pálsson (1991: 69) distinguishes analytically between two dimensions (i) *mode of circulation* with the possible values (a) *for use* and (b) *for exchange*, and (ii) *access to natural resources*, with the values (a) *non-ownership* and (b) *ownership*:

---

[3]That this has not always been the case has recently been demonstrated by Willson (2016) in her anthropological work on the forgotten seawomen of Iceland.

a. *Non-ownership/for use*: this production system is to be found in hunter-gatherer societies and is characterised by an egalitarian ethos and generalised reciprocity among the members of the community.

b. *Ownership/for use*: in this system of production access to resources is typically owned and distributed by certain clans or families with a long history and is inherited from generation to generation and legitimised through myths (for example, the 'first inhabitants' of a territory). Like in system (a), not much attention is paid to differences in success, but unlike the former system, the relationship between hunter and prey is not personal. Rather, and if acknowledged at all, success is considered a natural economic fact (Pálsson, 1991: 73–74).

c. *Non-ownership/for exchange*: This system of production is typical of market economies, where access to the resource base is open. In this system, labour is commodified and equipped with a certain accreditation, mostly in terms of power and of a personal and psychological nature that is more or less equivalent to luck (Pálsson, 1991: 75). Hence, in the 'hierarchical model', the status of a skipper is evaluated with regard to his catch, and, in contrast to models (a) and (b), his role is active.

d. *Ownership/for exchange*: this model is based on private ownership of the resource base in modern market economies where exclusive fishing rights are distributed by nation-states that rely on knowledge of scientific stock assessment. In this regime, capital and capture technology replace individual skill as explanatory variables to success (Pálsson, 1991: 77–78).[4]

In Icelandic public discourse, for instance, feudal metaphors such as 'quota kings' or 'lords of the sea' are used to describe the power relations of the new property regimes (Pálsson & Helgason, 1995, 1996).

---

[4]Pálsson and Durrenberger (1982, 1990) sparked a controversial academic debate by claiming that capital and fishing effort account for success in capitalist fisheries and not individual skill, as the folk myth of the 'skipper effect' suggests. However, authors such as Thorlindson (1988), Bjarnason and Thorlindson (1993), and Gatewood (1984) have convincingly demonstrated the remaining role of skill for success in contemporary fisheries. Accordingly, talented skippers will be rewarded with good positions on state of the art vessels, leading to a concentration of capital and skill.

Furthermore, Helgason and Pálsson (1997) have demonstrated how the commoditisation of fishing rights has led to an individualised discourse on 'greed' and 'laziness' as some quota-owners have taken advantage of leasing out shares of their quotas, which is known—in the eyes of those actually fishing—as 'fishing for others' (*veiða fyrir aðra*). Fishers and large parts of the public consider this form of profit-oriented exchange as immoral. Hence, the new property regime collides with folk accounts rooted in open-access to the resource base, in which those who put out to sea and fish are the rightful owners of their catch (Helgason & Pálsson, 1997: 458–459). In this sense, fishing rights are incommensurable for some participants in the Icelandic discourse, giving them a 'contested' status that has provoked a wide range of ongoing controversies that have been accompanied by acts of resistance in the form of disobedience and legal cases ranging from national courts up to the UN Court of Human Rights (Einarsson, 2011).

All in all, shifting the emphasis away from social actions illuminates the relations between regimes of production and societal discourse for understanding how representations of nature are constructed and manifested. The description of the architecture of these regimes and forms of coordination, however, remains rather vague. Moreover, we do not learn how the regimes of production are constructed and stabilised over time. The next section will try to shed some light on this matter.

The Cyborgisation of the Fisheries

In recent years, a number of sociologists have criticised the 'anthropocentric bias' in the social scientific understanding of modern resource management regimes that put the fisher at the centre of the understanding of fisheries dynamics or essentialise resource management regimes as 'ungovernable leviathan' (Holm, 2001, 2007; Holm & Nolde Nielsen, 2007; Johnsen, 2004, 2013; Johnsen, Murray, & Neis, 2009). Instead, so the claim goes, a *relational* understanding of modern resource management regimes is needed that takes into account the interrelations between the human and the non-human world.

In keeping with Actor–Network theory (ANT) (Latour, 2005; Law & Hassard, 1999), Holm (2001) rejects the Kantian divide between nature and culture/society that has led to the modernist struggle between naïve realism, which denies the power of symbols and ideas, on one hand, and radical constructivism, which downplays the role of the objective world, on the other. Instead, the success of the current fisheries management system is explained through the construction of 'reliable and stable translation chains' (Holm, 2001: 132) that mediate between the material and the symbolic world. Thus, the success of the Norwegian resource management regime can be explained neither through the naturalisation and taking for granted of norms in an institutional setting, nor through interest groups and power relations that shape the knowledge base of science. Moreover, the current resource regime is successful because it works in practice, as its actor-network was able to establish a stable and reliable alternative to the former regime.

To analyse these translation chains, Holm draws on ANT's principle of *extended symmetry* (Latour, 1999, 2005), which takes both human and non-human actors (fish, vessels, computers, storms and so on) into account and abolishes any division and hierarchy between the social and the natural, respectively. Accordingly, this principle is able to trace back the translation process that has led to the purification and naturalisation of scientific claims about fish, fish stocks and fishers, which in turn has ruled out the old network with a new stable and reliable network of nature–culture hybrids. At the heart of this 'invisible revolution' (Holm, 2001) lies the construction of Individual Transferable Quotas (ITQs). The most influential device in the resource management revolution is *Virtual Population Analysis* (VPA), which allows scientists to estimate the current state of fish stocks based on standardised samples in the oceans (Holm & Nolde Nielsen, 2007). In VPA, the condition of a fish stock for a specific cohort is predicted by the impact of the estimated 'natural mortality' (based on different factors that are very hard to predict with any accuracy) and 'fishery mortality' (based on landings). Even though VPA is far from operating under the conditions of an exact science, its policy impact was tremendous and key to the resource management revolution: 'With the stock assessment macroscope in place, the fisheries Leviathan could become powerful because he now could see and measure the object to be managed: the fish stock' (Holm & Nolde Nielsen, 2007: 179–180).

Even though a proposal for a fully-fledged ITQ-system was strongly opposed by fisher' organisations due to fears of quota concentration, centralisation and the possibility of a European takeover in 1991, fishing rights could be transferred on approval by the authorities. Ironically, the fishers themselves started showing an interest in trading quotas, which led to the development of an informal quota market. The introduction of the quota system led to a series of smaller reforms, which consolidated the system and reconstituted the fishers as economic and political agents with an interest in protecting and expanding their quota shares. Hence, also the fishers' organisation had to represent the interests of a group of privileged fish owners (Holm & Nolde Nielsen, 2007: 184–187). Thus, the construction of the Norwegian quota market can be viewed as an attempt to cool down the 'hot' controversies about overfishing and stock preservation that have emerged around the fish in question. Consequently, the *cyborg fish* has emerged as a powerful actor that 'goes hunting for the best owners' (Holm, 2007: 237–238). But has the cyborgisation of the fisheries moved modern resource management beyond the limits of governability?

In contrast to views that modern resource management ignores the social and ecological dynamics of fishery systems, Johnsen (2013: 2) has argued that 'governability is not limited by system properties but rather it is achieved through the objects and instruments that are deployed to make it possible'. Thus, ironically it is exactly its reductionism that has turned modern resource management into a successful governance regime. Inspired by Michel Foucault's (2008, 2009) inaugural lectures at the Collège de France, Johnsen outlines how the new resource management regime has developed from the modernist hierarchical system of direct control into a new form of governance based on indirect control and self-discipline, namely *neoliberal 'governmentality'*.[5] The notion of *governmentality* was coined by Foucault to indicate that the 'new art'

---

[5]Also see Lemke (2001) on the notion of 'neoliberal governmentality'.

of governance is based on discursively mediated and mentally self-imposed *techniques of the self*, which are indirectly stimulated by economic incentives based on monetarist reasoning and deregulation policies. In combination with an ANT perspective, which emphasises the artefacts and devices of this new art of fisheries governance, the research conducted on this theme clarifies how the reorganisation of fisheries governance has changed and how it affects the practices of harvesting. In this light, Johnsen, Murray, and Neis (2009) have described, on the basis of biographical interviews with fishers from northern Norway and Canada's east coast, how contemporary advances in modern resource management regimes have developed from *organic fisheries associations*, which are tightly embedded in the knowledge system of the coastal communities and characterised by informal ad hoc coordination and community ties, through hierarchical *mechanistic fisheries associations*, which can be characterised in terms of a high degree of differentiation, formalisation and professionalisation, to *cybernetic organisations* (Johnsen et al., 2009: 72). In contrast to the former ideal types, cybernetic fisheries organisations are described as 'techno-scientific systems', in which human and non-human relations are more tightly knit than in the previous regimes. As a consequence, this techno-scientific apparatus opens the way for new forms of coordination by new means of communication and feedback-based performance measures. In their most radical form, cybernetic fisheries organisations are based on the privatised distribution of fishing rights (ITQs), which tie fishing operations directly to the worlds of finance and change the relation between fish and fisher fundamentally. Hence, active members of the fishing industry must accept a certain set of rules and regulations to maintain their business under the auspices of formalised and quantifiable performance measures such as legal rules, technology and quality standards, medical tests, security checks and accounts of financial institutions (Johnsen et al., 2009: 73; Johnsen, 2013: 438–440). Within cybernetic organisations, however, power is only partly vertical as it is implicitly delegated 'in the work procedures on board and built into contractual property relations' (Johnsen et al., 2009: 70) by means of professionalisation and formalised roles. Thus, fishers become 'co-producers' of the blurred boundaries of the governing system and the system to be governed, respectively

(Johnsen, 2013: 14). Moreover, Johnsen (2004) has pointed out how financial incentives paired with technological innovations and increasing investments have replaced the typical small-scale, labour intensive and weather dependent fishing vessels with highly efficient 'capture machines' that have 'reduced the physical workload and standardised and increased the speed of operations' (ibid.: 487). As a result, today's resource management regime 'can be characterised as *integrated harvest machinery* within which fishers are not placed outside the machinery as users and masters, but must instead be seen as integrated parts of the machinery' (ibid.: 488, emphases in original). Thus, fishers have become increasingly dependent on technology that 'conveys values, goals and interests' (ibid.: 488). In this light, the failed policy of capture capacity reduction is the outcome of a bigger discourse on governance and technologies, rather than being an individualised 'tragedy of soft choices' (Standal & Aarset, 2002) on the part of rationally acting fishers.

Acknowledging the shifting focus to a fresh perspective that takes into account the role of materiality and science in the construction of markets, however, these accounts nevertheless remain rather vague regarding how the outcomes of these framing processes are structured and materialised in practice: how have markets changed daily coping and practice in market-based fisheries? Furthermore, these accounts remain blind to problems of coordination in what Polanyi (1957) has referred to as the 'empirical economy': to what extent have markets changed forms of production, valuation and exchange in rural production networks? And what are the consequences for coastal communities in this new culture of liberal rural capitalisms?

Conclusion: Fishing in Networks, Instutions and and Socio-Technical Discourse

Although contributions in the analysis of modern resource management in fisheries vary in scope and emphasises different questions and topics, it is easy to detect the lowest common denominator, which mainly revolves around a critique of the 'tragedy of the commons' narrative that serves as the basic paradigm of modern resource economics.

Accordingly, the models developed by resource economists neglect the social dimension of economic action and the endogenous nature of preferences that are shaped by networks, institutions, and socio-technical discourse, respectively. Accordingly, the methodological individualism deployed by economic models reduces social reality to the rational and efficient world of ego-centred and ahistorical homo oeconomicus, who can be governed by incentives, while downplaying the conflict potential of modern society. This review of the existing literature has shown that this image has been challenged from different perspectives that resemble prominent theoretical approaches in general social scientific research. Instead of seeing these accounts as mutual exclusive, we have seen how they complement each other and contribute to an encompassing sociological understanding of modern fisheries by highlighting the socio-material, cultural and institutional foundations in which they are anchored and reproduced.

With regards to this generally critical stance towards modern resource management, it is therefore not surprising that the respective literature is rich in promoting alternatives to modern resource management such as community-based governance (Acheson, 1988; Ostrom, 1990), co-management (Jentoft, 1989), local development (Phillipson & Symes, 2015; van de Walle, da Silva, O'Hara, & Soto, 2015) and small-scale production (Mackenzie, 2006; Shucksmith & Rønningen, 2011). A similar critical stance can be found in more general approaches, which scrutinise the role of globalisation and neoliberal reform in the countryside around issues such as the role of powerful macro-actors (Friedmann & McMichael, 1989; McMichael, 2005), knowledge regimes (Busch, 2007, 2010), the socio-technical infrastructure of agricultural markets (Çalişkan, 2010; Garcia-Parpet, 2008) and the role of rural entrepreneurs for institutional change (Barbera & Audifredi, 2012), respectively.[6]

While these accounts likewise provided valuable insights into the role of markets for rural regions, it is nevertheless surprising that little

---

[6]For a thorough review of the literature on markets in rural economies, see Dobeson (2018).

has been said about how the privatisation of common resources and the organisation of new markets for access rights has transformed the cultural economy of the countryside as such. Instead, it seems as social scientists have left the understanding of markets to resource economists, who appear to be quite successful with promoting their hegemonic vision of markets as the most efficient means for environmental protection and economic growth. In order to understand how new markets, property rights and technology transform rural economies, the next chapters will therefore unravel the social relations, institutions and discourses that make up the world of fishing by zooming in on the daily practical coping of fishers with an increasingly economised and globally entangled world. We will see that Bjartur's metamorphosis has to be understood in the broader context of a comprehensive economisation of production, valuation and exchange, which is first and foremost to be understood as a result of the changing practices and expectations about the future that have come along with the privatisation of access rights.

References

Acheson, J. M. (1977). Technical Skills and Success in the Maine Lobster Industry. *Human Ecology, 3*(3), 183–207.

Acheson, J. M. (1981). Anthropology of Fishing. *Annual Reviews of Anthropology, 3*(3), 275–316.

Acheson, J. M. (1988). *The Lobster Gangs of Maine.* Hanover: University Press of New England.

Acheson, J. M. (1998). Lobster Trap Limits: A Solution to a Communal Action Problem. *Human Organization, 57*(1), 43–52.

Apostle, R., Barret, G., Holm, P., Jentoft, S., Mazany, L., McCay, B., & Mikaelsen, K. (1998). *Community, State and Market on the North Atlantic Rim: Challenges to Modernity in the Fisheries.* Toronto: University of Toronto Press.

Arnason, R., & Runolfsson, B. T. (Eds.). (2008). *Advances in Rights Based Fishing: Extending the Role of Property in Fisheries Management.* Reykjavík: Bókafélagið Ugla.

Aspers, P. (2011). *Markets*. Cambridge: Polity Press.

Barbera, F., & Audifredi, S. (2012). In Pursuit of Quality. The Institutional Change of Wine Production Market in Piedmont. *Sociologia Ruralis, 52*(3), 311–331.

Barnes, J. A. (1954). Class and Commitees in a Norwegian Island Parish. *Human Relations, 7*, 39–58.

Bjarnason, T., & Thorlindson, T. (1993). In Defense of a Folk Model: The "Skipper Effect" in the Icelandic Cod Fishery. *American Anthropologist, 95*(2), 371–394.

Busch, L. (2007). Performing the Economy, Performing Science: From Neoclassical to Supply Chain Models in the Agrifood Sector. *Economy and Society, 36*(3), 437–466.

Busch, L. (2010). Can Fairy Tales Come True? The Surprising Story of Neoliberalism and World Agriculture. *Sociologia Ruralis, 50*(4), 331–351.

Çalişkan, K. (2010). *Market Threads: How Cotton Farmers and Traders Create a Global Market*. Princeton: Princeton University Press.

Cardwell, E. (2015). Power and Performativity in the Creation of the UK Fishing-Rights Market. *Journal of Cultural Economy, 8*(6), 705–720.

Dobeson, A. (2018). Economising the Rural: How New Markets and Technology Transform Rural Economies. *Sociologia Ruralis, 58*(4), 886–908.

Einarsson, N. (2011). Fisheries Governance and Social Discourse in Post-crisis Iceland: Responses to the UN Human Rights Committee's View in Case 1306/2004. *Yearbook of Polar Law, 3*, 470–515.

European Commission. (2009). *The Common Fisheries Policy: A Users Guide*. Luxembourg: Office for Official Publications of the European Commission.

Eythórsson, E. (1996). Theory and Practice of ITQs in Iceland. *Marine Policy, 20*(3), 269–281.

Eythórsson, E. (2000). A Decade of ITQ-Management in Icelandic Fisheries: Consolidation Without Consensus. *Marine Policy, 24*, 483–492.

Foucault, M. (2008). *Security, Territory, Population: Lectures at the Collège de France 1978–1979*. Basingstoke: Palgrave Macmillan.

Foucault, M. (2009). *The Birth of Biopolitics: Lectures at the Collège de France 1978–1979*. Basingstoke: Palgrave Macmillan.

Friedmann, H., & McMichael, P. (1989). Agriculture and the State System: The Rise and Decline of National Agricultures, 1870 to the Present. *Sociologia Ruralis, XXIX*(2), 93–117.

Garcia-Parpet, M.-F. (2008). The Social Construction of a Perfect Market: The Strawberry Auction at Fontaines-en-Sologne. In *Do Economists Make Markets? On the Performativity of Economics* (pp. 20–53). Princeton: Princeton University Press.

Gatewood, J. B. (1984). Is the "Skipper Effect" Really a False Ideology? *American Ethnologist, 11*(2), 378–379.

Granovetter, M. (1985). Economic Action and Social Structure: The Problem of Embeddedness. *American Journal of Sociology, 40*(3), 481–510.

Hannesson, R. (2004). *The Privatization of the Oceans*. Cambridge, MA and London, UK: MIT Press.

Hardin, G. (1968). The Tragedy of the Commons. *Science, 162*, 1243–1248.

Harvey, D. (2005). *A Brief History of Neoliberalism*. Oxford: Oxford University Press.

Helgason, A., & Pálsson, G. (1997). Contested Commodities: The Moral Landscape of Modernist Regimes. *The Journal of the Royal Anthropological Institute, 3*(3), 451–471.

Hersoug, B. (1997). What is Good for the Fishermen, is Good for the Nation: Co-management in the Norwegian Fishing Industry in the 1990s. *Ocean & Coastal Management, 35*(2–33), 157–172.

Hersoug, B. (2002). *Unfinished Business: New Zealand's Experience with Rights-Based Fisheries Management*. Delft: Eburon.

Hersoug, B. (2005). *Closing the Commons: Norwegian Fisheries from Open Access to Private Property*. Delft: Eburon.

Hersoug, B., Holm, P., & Rånes, S. A. (2000). The Missing T. Path Dependency Within an Individual Vessel Quota System—The Case of the Norwegian cod Fisheries. *Marine Policy, 24*, 319–330.

Hollingsworth, R. J., & Boyer, R. (Eds.). (1997). *Contemporary Capitalism: The Embeddedness of Institutions*. Cambridge: Cambridge University Press.

Holm, P. (2001). *The Invisible Revolution: The Construction of Institutional Change in the Fisheries* (PhD dissertation). University of Tromsø, Tromsø.

Holm, P. (2007). Which Way Is Up on Callon? In F. Muniesa, L. Siu, & D. MacKenzie (Eds.), *Do Economists Make Markets?* (pp. 225–243). Princeton: Princeton University Press.

Holm, P., & Nolde Nielsen, K. (2007). Framing Fish, Making Markets: The Construction of Individual Transferable Quotas. In Y. Millo, M. Callon, & F. Muniesa (Eds.), *Market Devices* (Vol. 55, pp. 173–195). Malden: Blackwell.

Høst, J. (2015). *Market-Based Fisheries Management: Private Fish and Captains of Finance*. Dordrecht: Springer.

Jentoft, S. (1989). Fisheries Co-management. *Marine Policy, 13*(2), 137–154.

Johnsen, J. P. (2004). The Evolution of the "Harvest Machinery": Why Capture Capacity Has Continued to Expand in Norwegian Fisheries. *Marine Policy, 29*, 481–493.

Johnsen, J. P. (2013). Is Fisheries Governance Possible? *Fish and Fisheries, 15*(3), 428–444.

Johnsen, J. P., Murray, G. D., & Neis, B. (2009). North Atlantic Fisheries in Change: From Organic Associations to Cybernetic Organizations. *Mast, 7*(2), 55–82.

Keohane, N. O., & Olmstead, S. (2007). *Markets and the Environment.* Washington, DC: Island Press.

Latour, B. (1999). *Pandora's Hope: Essays on the Reality of Science Studies.* Cambridge, MA: Harvard University Press.

Latour, B. (2005). *Reassembling the Social: An Introduction to Actor-Network— Theory.* Oxford: Oxford University Press.

Law, J., & Hassard, J. (1999). *Actor Network Theory and After.* Oxford: Blackwell.

Lemke, T. (2001). The Birth of Bio-Politics. Michel Foucault's Lecture at the Collège de France on Neo-Liberal Governmentality. *Economy and Society, 30,* 190–207.

Mackenzie, A. F. (2006). A Working Land: Crofting Communities, Place and the Politics of the Possible in Post-Land Reform Scotland. *Transactions of the Institute of British Geographers, 31*(3), 383–398.

Macneil, I. (1978). Contracts: Adjustment of Long-Term Economic Relations Under Classical, Neoclassical and Relational Contract Law. *Northwestern University Law Review, 72,* 854–887.

McMichael, P. (2005). Global Development and the Corporate Food Regime. In P. D. McMichael & F. H. Buttel (Eds.), *New Directions in the Sociology of Global Development* (Vol. 11, pp. 269–303). Bingley: Emerald.

Miller, M. L., & Van Maanen, J. (1979). "Boats Don't Fish, People Do": Some Ethnographic Notes on the Federal Management of Fisheries in Gloucester. *Human Organization, 38*(4), 377–385.

Miller, M. L., & Van Maanen, J. (1982). Getting Into Fishing: Observations on the Social Identities of New England Fishermen. *Journal of Contemporary Ethnography, 11*(1), 27–54.

Ostrom, E. (1990). *Governing the Commons: The Evolution of Institutions for Collective Action.* Cambridge: Cambridge University Press.

Pálsson, G. (1991). *Coastal Economies, Cultural Accounts: Human Ecology and Icelandic Discourse.* Manchester: Manchester University Press.

Pálsson, G., & Durrenberger, E. P. (1982). To Dream of Fish: The Causes of Icelandic Skipper's Fishing Success. *Journal of Anthropological Research, 38*(2), 227–242.

Pálsson, G., & Durrenberger, E. P. (1990). Systems of Production and Social Discourse: The Skipper Effect Revisited. *American Anthropologist, 92*(1), 130–141.

Pálsson, G., & Helgason, A. (1995). Figuring Fish and Measuring Men: The Individual Transerable Quota System in the Icelandic Cod Fishery. *Ocean and Coastal Management, 28*(1–3), 117–146.

Pálsson, G., & Helgason, A. (1996). The Politics of Production: Enclosure, Equity, and Efficiency. In G. Pálsson & E. P. Durrenberger (Eds.), *Images of Contemporary Iceland: Everyday Lives and Global Contexts* (pp. 60–86). Iowa City: University of Iowa Press.

Phillipson, J., & Symes, D. (2015). Finding a Middle Way to Develop Europe's Fisheries Dependent Areas: The Role of Fisheries Local Action Groups. *Sociologia Ruralis, 55*(3), 343–359.

Polanyi, K. (1957). The Economy as Instituted Process. In M. Granovetter & R. Swedberg (Eds.), *The Sociology of Economic Life* (pp. 31–50). Cambridge: Westview Press.

Polanyi, K. (2001). *The Great Transformation: The Political and Economic Origins of Our Time*. Boston: Beacon.

Shucksmith, M., & Rønningen, K. (2011). The Uplands After Neoliberalism?—The Role of the Small Farm in Rural Sustainability. *Journal of Rural Studies, 27*, 275–287.

Skaptadóttir, U. D. (1996). Housework and Wage Work: Gender in Icelandic Fishing Communities. In G. Pálsson & E. P. Durrenberger (Eds.), *Images of Contemporary Iceland: Everyday Lives and Global Contexts* (pp. 87–105). Iowa: University of Iowa Press.

Standal, D., & Aarset, B. (2002). The Tragedy of Soft Choices: Capacity Accumulation and Lopsided Allocation in the Norwegian Coastal Cod Fishery. *Marine Policy, 26*, 221–230.

Svallfors, S., & Tyllström, A. (2018). Resilient Privatization: The Puzzling Case of for-Profit Welfare Providers in Sweden. *Socio-Economic Review*. Retrieved from https://doi.org/10.1093/ser/mwy005.

Sverisson, Á. (2002). Small Boats and Large Ships. Social Continuity and Technical Change in the Icelandic Fisheries, 1800–1960. *Technology and Culture, 43*(2), 227–253.

van de Walle, G., da Silva, S. G., O'Hara, E., & Soto, P. (2015). Achieving Sustainable Development of Local Fishing Interests: The Case of Pays d'Auray Flag. *Sociologia Ruralis, 55*(3), 360–377.

Williamson, O. (1975). *Markets and Hierarchies: Analysis and Anti-trust Implications: A Study in the Economics of Internal Organization*. New York: Free Press.

Willson, M. (2016). *Seawomen of Iceland: Survival on the Edge*. Seattle and London: University of Washington Press.

3

Economising Rural Independence

> *I think these small boats are mainly about*
> *being independent; if that was not the main*
> *reason, then I think there will be no more boats*
>
> Skipper

A life in the harbour

Traditionally, being a small-boat fisher in Iceland is not about profit-making or living an outlaw lifestyle apart from society. Rather, small boats embody a culture of living together with others in an independent community in which a fleet of vessels contributes to the local economy, as Bjartur, our independent small boat fisher explains:

> And this is what we've been fighting for [the small boats], because the small boats we say, it's a life in the harbour – can you imagine how it is to have 12 boats? All boats go out in the morning and come in the same night, then it's a lot to do in the harbour. But when you're alone, then there's nothing much happening – you know what I mean? (MII)

Hence, a thriving small-boat port is constantly in motion and provides reliable work for the community in contrast to harbours with one or

© The Author(s) 2019, corrected publication 2020
A. Dobeson, *Revaluing Coastal Fisheries*,
https://doi.org/10.1007/978-3-030-05087-0_3

two bigger vessels that return to port every few weeks; being independent implies a form of collective solidarity that provides work and stable income for the members of the community.

Although many coastal communities had their heyday with the trawler fisheries, they were also a source of instability and bankruptcies for many of them. In this sense, small boats have long built a safety net for many coastal communities, which provided relative flexibility and stability in times of economic recession. In this context, Bjartur continues to remark that 'it's also that you can live in place like this, you can't rely always on the big companies' (ibid.)—and he certainly knows what he is talking about, as the big trawler company that used to provide the economic backbone of the community decided to sell the company overnight just before 2008 recession, leaving dozens of locals who previously worked on the trawler or in processing and services without a job from one day to another. And even today, the fear of being dependent on a single quota owner is widespread in many coastal communities in which bigger companies dominate the local economy.

Being independent on a small boat is seen not only as a means of profit-making, but also as a form of resilience and rural independence. In this regard another skipper highlights the labour-intensive and artisanal aspect of the coastal fisheries as important pillars of the local economy:

> I believe that it's a culture for our business to have a small boat like this! And it also makes much work for people, so we need a lot of people to work with and that's good for Iceland, it's much better for Iceland to have all the people in work than unemployment. (...) I believe that we don't need this technique to do it, a lot of things we can do by hand, and I think that's good! Everyone has, must have something to do, that's much better for Iceland overall! (XIX)

Put differently, small boats have not meant to be primarily directed towards economic efficiency and profit-making, as they are 'all about independence because they own these boats free from these big companies!' (MVII). Moreover, small boats are cultural signifiers representing 'part of a lifestyle' (MXX), a tradition that is passed on from generation to generation and that is not mainly oriented towards making a profit and driving economic growth, as an old veteran recapitulates:

> Making a profit has never been the main thing. The family tradition of having a boat, fishing for your own consumption and pleasure, being in direct contact with nature and being able to charge your own batteries for the winter has been more important. (MXIV)

For him, as well as for many others living in remote communities, small boats signify more of an intrinsic value of living a life in close communion with nature and maintaining the life of a community. Today, however, it seems as that the traditional meaning of small boats as a symbol of rural independence is increasingly being challenged by a new set of values and practices emerging due to the reorganisation of the fishing industry around new markets and property rights. Although many coastal fishers in the rural regions were against the quota system in fear that they could not compete with the capital-strong companies around Reykjavik, some of the original protesters find themselves belonging to the exclusive class of quota-owners that is much detested by the political left. In other words, Bjartur, as many other small boat fishers seem to have become 'one of them', leading many people to believe that they have betrayed their cultural heritage in the name of money-making. Many small-boat owners, however, think that these accusations are based on the ignorance of many people of the reality of life for small-boat owners when the quota system was implemented. As Bjartur's mother Krístin, a former fisher and shareholder in the company, remarks:

> I was standing with a fire [torch] in front of the parliament [protesting against the quota system]. But now we bought quotas, took loans and so. People do not understand. It was a decision between life and death. We didn't want to be slaves and move to Reykjavík. We are too old to get jobs there. (FN: 71)

It becomes clear that the market-based reform has left the family with a tough decision: either sell off the allotted quota-share or to betray their own ideal to remain their dream as independent small boat fishers. In order to understand the consequences of playing along with the quota system for the culture of small boats, this chapter attempts to reconstruct and explain the transformation of small boats from symbols and means of rural independence to objects of investments and profit-making. By drawing on the economic sociology literature on economisation and marketisation we will see how new markets and property rights has fundamentally transformed the lifeworld of Bjartur and his fellows.

The chapter is structured as follows. First, a brief historical sketch illustrates the political struggle that has shaped market institutions and successively disentangled small boat owners from their rural ties to becoming truly independent and 'free' market actors in the economic sense. Subsequently, the reader will witness how this new class of independent small boat fishers has become successively *re-entangled* and hooked into a new network of money-mediated expectations that has not only transformed the practices, but also the material culture of coastal fisheries as such.

Disentangling Small Boats

Ever since the dawn of modernity, large-scale industrial capitalism has been the dominant institutional arrangement in many Western fisheries, which has transformed and 'disembedded' coastal communities from their community ties (Apostle et al., 1998). Although some communities in the North Atlantic managed to re-embed their local economies, a shift towards market-based models of resource management marks the new regime of neoliberal fisheries management. New markets and property rights, however, do not come out of the blue. Nor are they the necessary outcome of some sort of evolutionary societal development. Rather, markets, as for fishing quotas, are *made* (Aspers, 2009) and organised within historically-grounded, often contested fields of power that define and limit the rules and boundaries of valuation and exchange (Ahrne, Aspers, & Brunsson, 2015; Fligstein, 1996). Scholars within the field of Science and Technology Studies have gone even further by deconstructing the very concept of 'the economy' and 'the market' as such (Callon, 1998). Instead of simply taking actors and social structures for granted, Çalişkan and Callon (2009, 2010) highlight the socio-technical discourses and practices that allow the 'disentanglement' and re-shaping of actors and objects into economic entities as perceived by social scientists and economists. From this perspective, economisation and marketisation can rather be understood as a socio-technical process of *translation* (Callon, 1986) of a given an actor-network such as the Icelandic fisheries into the language of the neoclassical market model. This translation process, in modern fisheries management provided by resource economists, thus provides the backbone of a process of *framing*, in which so-called 'calculative agents' including fishers and fish stocks are *disentangled* from the complexities of the empirical world and isolated from the externalities of decision-making (Callon, 1999). In this sense, the neoclassical textbook model of homo oeconomicus is neither a fiction, nor does it describe the essence of man, but is a model according to which reality is created by means of commensuration into standardised metrics that allow for rational decision-making. From this perspective, the ideal world of the neoclassical market model in which demand clears supply based on mutual adjustment of rational calculating individuals is not 'real' in itself, but constructed and performed through 'market devices' such as modern resource economics and scientific stock assessment that enable calculative agency to come about (Callon, Millo, & Muniesa, 2007). Hence, economisation is not merely a structural force of society, but a set of socio-technical practices and devices that altogether co-constitute the world as 'economic' (Çalişkan & Callon, 2010: 2). In other words, the interesting aspect for our understanding of economisation and marketisation is not that the neoclassical market model is empirically wrong, but that it has reconfigured the world of modern fisheries as such. But what does this new regime of market-based governance imply for the daily practices of Bjartur and his fellow small boat fishers?

The following section will reconstruct how the organisation of new markets and property rights in the have successively *disentangled* small boat fishers from their rural ties while at the same time reconfiguring them as 'free' entrepreneurial market actors who coordinate their fishing operations with regard to market movements. At the same time, however, the opportunities that come with the market system have successively *re-entangled* small boat fishers in a complex money-mediated web of multiple market structures, banks, global investments, scientific discourse and political decision-making that pushes them to economise their expectations toward increasing profit-making to stay afloat. The consequences of this are paradoxical: while the economisation of the small-boat economy has disentangled the rural small-boat fishers from their local ties enables Bjartur and his fellows to maintain their livelihoods as relatively independent small boat fishers, the cultural value of small boats as symbol of rural independence is being challenged by a set of new expectations resulting from the new entanglements and practices in a globalised world of volatile markets, debts and capital flows.

From Dead Fish to Lively Capital

Although fishing did not form the bedrock of Iceland's prosperity and wealth until the industrialisation of the fisheries in the late nineteenth century, numerous coastal communities flourished and became dependent on the fisheries as the primary source of income with the advent motor by the turn of the century. 'Herring towns' such as Siglufjörður in north Iceland emerged in the 1940s and 1950s and attracted a lot of people to work in the fishing industry, both seasonally and permanently. Even though this period is referred to as 'the good old days' among Icelanders, competition between fishers was harsh as more and more vessels gathered around the Icelandic shoreline, trying to make a living from the rugged, but highly profitable Icelandic waters. Most of the coastal communities, however, had to participate in direct competition not only with their own people, but also with foreign fishing vessels, most notably from Britain and West Germany. At the same time, new technologies fostered a rapid increase in fishing efforts and the emergence of new mass markets put a lot of pressure on the stocks. Not only were the fishing communities aware early on that the foundation of their wealth was not infinite, but also government authorities started to realise that the fish stocks have to be protected if the nation's most important source of prosperity and wealth was to be maintained.

To protect this wealth, the government made great efforts to limit access to its fishing grounds by the steady expansion of an Exclusive Economic Zone (EEZ). The implementation of the 12-mile zone marked the starting point to the notorious 'cod wars' between Iceland and the UK, Germany and Belgium—who have been fishing off the Icelandic shore for centuries—in 1958. After the conflict was resolved, foreign vessels were allowed to maintain their fishing activities for a few more years (Ingimundarson, 2008). But after the few rather benign incidents between the Icelandic coast guard and British and German vessels in 1958, the conflict heated up again after Iceland declared the expansion of its EEZ from 12 to 50 nm in 1972 and from 50 to 200 nm in 1975. The so-called second and third cod wars were accompanied by net-cutting and serious ramming incidents between the Icelandic coastguard and British fishing vessels that continued their fishing activity within the Icelandic EEZ, and boycotts of Icelandic produce in the UK's domestic market. This long-lasting conflict was not fully resolved until 1976, when the Icelandic government threatened to close down a strategically important NATO base in Keflavik in the midst of the Cold War if the expansion of the nation's EEZ to 200 nautical miles was not accepted. Even though Iceland, backed by the United States, finally prevailed, the 200 nm zone was not enough to mitigate catch efforts and control fish stocks effectively as the national fleet continued to expand its catch capacity and fishing efforts steadily.

As a consequence, the herring stocks collapsed due to the fishing boom in 1968 and marked the starting point of systematic attempts to reduce catch capacity by the implementation of Total Allowable Catches (TACs), which were first introduced for herring (1969) and supplemented by a TAC for cod in 1976, which was to become the new anchor currency of the Icelandic fisheries. These restrictions, however, did not prevent fish stocks from overfishing as fleet capacity increased steadily and finally turned out to be far too big for the total quota, and hence shifted its fishing efforts to other species, notably cod. The fishers engaged in so-called 'Olympic fishing', and slowly but surely, it seemed that history was about to repeat itself in the early 1980s.

In direct response to the Marine Research Institute's (Hafrannsóknastofnunin) 'Black Report' in 1983, in which the poor condition of the cod stock was announced, the government took the initiative to prevent the cod fisheries from the fatal consequences of the herring bonanza by implementing a vessel quota in the demersal fisheries in 1984, which represents the starting point and bedrock of the current Icelandic resource management system. Initially, the idea of vessel quotas was thought of as an 'experiment', which was to last for at least one year.

In the beginning, the system gained much support among boat owners and was only opposed by a minority from the Officer's Union (FFSÍ) and regional representatives of small rural fishing communities, such as the Westfjords, who in the end accepted the trial at least until the cod stock had recovered again (Eythórsson, 2000: 485). The outcome of this crisis management was the *Fisheries Management Act* (FMA) of 1983, which entitled any boat that had been fishing Icelandic waters for the previous three years to a quota share based on the catch history of each individual vessel. Alternatively, boat owners could also choose to fish within a system of 'effort quotas', which was an option especially for those who have not been fishing much over the previous three years.

Small boats under 10 tonnes could continue their operations without being subject to either of the two systems in the beginning, but were integrated into the quota system in 1985 by a special kind of effort quota, which made small boats extremely popular among boat owners. According to (Eythórsson, 2000: 486) 'it might seem that the quota system would "wither away", as a majority of boat owners opted for the effort-quota alternative in order to increase their share of the TAC at the expense of those regulated by ITQs'. Hence, the small-boat fleet increased tremendously[1] as a more liberal management regime and improvements in fishing gear appeared to be very attractive to many boat owners. As a consequence, the system failed as fleet size and catch capacity increased continuously. These developments called for a

---

[1] 'While 964 small boats were registered in 1984, their number had increased to 1956 in 1990' (see Eythórsson, 2000: 486).

revision of the FMA in the period 1988–1989. As a result, the FMA of 1990 abolished the effort quota system with the exception of small boats up to 6GRT and turned vessel quotas into permanent and freely transferable property rights.

The disentanglement of the fishers and their translation into quota-owners in the sense of modern resource management went hand in hand with the consolidation of the quota system. Similar to the discourse on emission markets (MacKenzie, 2009), the quota system made it possible to cool down controversies around environmental discourses that not only included environmental interests, but also the fishers by opening up new business opportunities for those with privileged access to it. Due to the permanent status and potentially increasing value, quotas could be used as collateral, which led to investments that soon exceeded the total value of the fishing industry. As Hannes Hólmsteinn Gissuaarson, one of the main ideologues of the political right stated in *Ísland i dag* at the peak of the financial bubble:

> We activate capital that was previously dead … The fish stocks did not have a price tag, they were non-transferable and could not be used as collateral – non-tradable. Then the quotas are allocated, which creates capital … Here in Iceland, capital was handed over to private owners, and then it became alive. (cited after Benediktsson & Karlsdóttir, 2011: 231)

It becomes clear that although fishing quotas were first implemented as conservation measure, market liberals soon discovered the potential of turning dead fish into lively capital in the sea. Hence, as long as access to the sea is restricted and fish stocks remain on a more or less sustainable level, the booming demand for cheap protein will let the value of fishing rights soar, just like the booming real-estate market in the capital region. In other words, ITQs bear the potential to create 'new wealth', as one of the academic fathers of the Icelandic quota system claimed in an article published just before the financial meltdown of the Icelandic economy (Arnason, 2008). Accordingly, the mobilisation of lively capital from the sea has not only triggered unseen investments and rationalisation within the fishing industry, but even created new value in other parts of the booming Icelandic economy and abroad. By unfolding its potential as new form of bio-capital, ITQs have become the symbol and paragon of the new era of financial capitalism and aggressive expansion of the Icelandic economy for the political right. In a way, the activation of dead capital in the sea by putting a price on its future availability spearheaded the policy reforms of decentralisation and financialisation that transformed the hitherto still rather centralised and closed national economy.

The mobilisation of lively capital, however, was not friction-free, as it sparked public controversies and outrage about the ownership of the nation's most valuable resource. Early critiques of the quota system soon came to highlight the injustice the system has created with regard to the concentration of fishing rights in the hands of a small number of 'quota-kings' (Helgason & Pálsson, 1997).[2] Until the present day, the ITQ-system remains to be the contested subject of fierce and highly emotional controversies about the right to fish which is deeply ingrained in the collective identity of Iceland as an independent people. Ironically, it is §1 of the FMA itself stating that

> the exploitable marine stocks of the Icelandic fishing banks are the common property of the Icelandic nation. (Fiskistofa, 2006)

No cultural artefact symbolises this 'right to fish' more than the small boat fisheries, which have played an important economic role since the early settlements (Þór, 2002) and survived the modernisation of the fishing fleet by counterbalancing the risks of large-scale industrialism through gradual technological adaption (Sverisson, 2002). The belief in being free to access the nation's fishing grounds has even led some small boat fishers convicted of illegal fishing to bring the issue in front of the UN Human Rights Committee, which has condemned the closure of the marine commons (Einarsson, 2011). Despite this criticism, however, the ITQ-system has shown itself rather resilient against political reform. Instead of simply washing all small boats away, however, we will

---

[2]The largest ten quota holders are in control of over 50% of the total quota (Fiskistofa, 2012, author's calculations).

see how mobilisation of capital has turned the protest of the small boat owners itself into a new source of value generation, as the following will show.

From Resilient Peasantry to Independent Market Actors

With implementation of the ITQ system in 1984, many inhabitants of rural regions along the coast were afraid that their right to fish as independent fishers would be contested through the closure and capitalisation of the fishing industry. In particular, fishers from the Westfjords, which traditionally held the bulk of the coastal fleet, feared that the quota system would destroy the foundations of livelihood and a centuries-old tradition, as they saw capital-strong investors from the Reykjavik region buying up quotas from rural communities or selling out of the industry and consequently weakening the coastal communities. This cultural meaning was reason enough for many small-boat fishers, who saw the ITQ system as a threat to their autonomy to live and work in solidarity with each other. As a result, the National Association of Small Boat Owners (NASBO) was founded in 1985 as a protest movement against the quota system and has become a strong oppositional force against large-scale corporate activities and fishing restrictions and a lobbying organisation representing small-boat owners in parliament and the media.[3] In this role, NASBO engages in political mobilisation and manages to maintain the autonomous status of the coastal fisheries as different from the large-scale industrial fleet. To protect coastal fishers, access was first kept open for all coastal fishing vessels. Due to increasing lobbying from the LÍU (Icelandic Federation of Fishing Vessel Owners) and the discursive shift of policymakers towards property-rights-based management solutions (Eythórsson, 2000), small boats were successively translated into carriers of exclusive fishing quotas. First, quotas were allotted based on a vessel's individual fishing catch

---

[3]According to NASBO (LS 2011) their members owned 1150 vessels in 2010/2011, which makes about 70% of the total fleet in 2010 (Statistics Iceland, 2011).

records from the past three years in 1991. In this period, fishers could choose between fishing days at sea or ITQs. With lower TACs and rising prices for quotas, more and more fishers decided to opt for quotas over the years, as days at sea got fewer and fewer. In 2004, all remaining fishers were transferred into a separate small boat ITQ-system, marking the final stages of the metamorphosis of independent peasantry to relatively independent market actors.

As of now, all small boats have been successively disentangled from their homeport communities and translated into vehicles attached to highly valuable assets. The same holds true for remaining fishers, who were literally turned over night into privileged owner of the nation's most valuable natural resource. One could say that Bjartur and his fellows were forced to being truly independent, although often against the understanding of other community members or the 'townspeople' from the capital region. 'We didn't ask for it [the quota]!', as Bjartur's mother remarks, and no matter how the family would have decided they would have probably been criticised. But it is clear that the possibility of becoming a quota owner also indicated a new opportunity for the family to maintain their livelihoods as independent small boat fishers. The decision to stay in the industry, however, did not mean that the family could simply rest on its new status as independent market actors with privileged access to the nation's symbol of wealth and prosperity. Rather, they were now confronted with a complex set of new rules and regulations, which changed the expectations of what it means to be independent.

The Architecture of the Small Boat Economy

The rise of ITQs in Iceland shows yet again that markets do not simply result as the spontaneous coming together of individual actors, but rather result from highly contested power struggles between different interest groups that altogether shape the institutional 'architecture' of market organisation (Fligstein, 2002). To the present day, the struggle around the right is reflected in the complex architecture of the small boat economy, which has been widely closed in terms of access. At the same time, however, political pressure from small boat owners has led to the institutionalisation of a set of rules and regulations ought to protect small boats from their large-scale counterparts. Within the small-boat ITQ system, this has led to further differentiation into three different licensing systems based on vessel size and fishing gear:

a. The small-boat *ITQ system* including vessels of up to 10 GT, which are allowed to use any kind of legal fishing gear.[4]
b. The Hook-and-line (Krókaaflamarkskerfi, H&L) ITQ system for vessels of up to 15 GT (changed to 30 GT in 2013), which are restricted to the use of hook-based capture technology, such as longlines and jigging (handfæri).[5],[6]

In addition, vessels smaller than 12 MT can hold a special *lumpfish license*, which is limited to 400 in total and 50 days at sea for each licence. Lumpfish are usually caught near the coastal region with gill nets and its precious roe and caviar and makes an important economic contribution to many small-boat fishers with low quotas.

In principle, ITQs can be transferred freely in the small-boat ITQ system and from the large-scale system to the small-scale system. In order to protect small boats, however, transfers from the small-boat system to the large-scale system are prohibited. Moreover, quota-owners in the small-boat ITQ system must not exceed the limit of owning more than 4% of the total H&L share for cod and 5% of total share for haddock or not exceed 12% of the total value of the catch share to prevent the centralisation of fishing rights and monopolies (cf. Fiskistofa, 2006: Articles 12–13). Furthermore, quota-owners must land at least 50% of their quota share at their homeport over two years in order to maintain their rights. If this obligation is not fulfilled, the owner will be expropriated without compensation and quotas allotted among other quota holders accordingly (Fiskistofa, 2006: Article 15).

---

[4]In 2012, 215 vessels were registered in the small-boat ITQ system, contributing to about 12% of total landings in the small-boat sector (Þórðarson & Viðarsson, 2014: 7).

[5]See appendix for a detailed description of fishing gear; see also Chapter 5.

[6]The H&L system accounts for the bulk of catches in the small-boat system, with 77% of total landings in the small-boat sector in 2012 (Þórðarson & Viðarsson, 2014: 7).

On top of these protective legal measures, the government has created some economic incentives to maintain small boats as the backbone of rural community development, most importantly the 'quota discount' for hand instead of machine baited lines to maintain jobs in the community (see Appendix).[7] In addition to this, fishing communities in decline may apply for a special 'community quota', which is allocated to local fishers, who are not allowed to trade these allotted quota and must land their catch within the same community for processing.

The Fish Auction

Another important pillar of the architecture of the small boat economy that contributes to the disentanglement of small boats and fishers from their rural ties is the organisation of a national fish auction for fresh fish. Going hand in hand with the privatisation of fishing rights, the collective organisation of fish auctions in the late 1980s was previously seen as an attempt to attract fishing vessels to land their catch in struggling communities (Graham, 1998). With advancements in digital information technology that allows combine all local fish auctions into one encompassing national and fully digitised auction system, the old community-bound network of strong ties has been disentangled by the organisation of a new market, which allows fishers to coordinate their fishing activities with regard to fluctuating prices, rather than being loyal to or forced to accept the local prices of processors in their homeport communities.

As a consequence of the disentanglement of production and fishing, most independent fishers choose to sell their catch on the auction market, which is usually more lucrative than being contracted at a fixed price.[8] This price autonomy is generally perceived as a positive

---

[7]Accordingly, small boats that land their catch within 24 hours at their homeport community may land 16% (20% in 2014) in excess of their fishing quota if the fishing line was hand-baited in the same community (Fiskistofa, 2006: Article 11).

[8]Older fishers with long-lasting relations, or quota owners in poorer communities fishing on subsidised 'community quotas', however, may still have fixed arrangements with local processors, and vessels owned by processors of course deliver most of their fish to the company's plant, although by-catch or species not suitable for production are usually sold off to the auction market.

development by many fishers and even considered an integral part of the resilience and modernisation of the small-boat fisheries, as a small boat fisher claims:

> What I can say about the fish markets or the auction market is that this is one of the best things that have happened to the small boats because before they were only selling their catch directly to the processors and the processors just had their way with the pricing, they could only do this because of, you know, lack of good communications, I mean with the old system and that. But with the auction market this has started to turn the tide and today this is really one of the main factors why the small boats survive: they get the highest price through the auction market. (MV)

Field observations suggest that some boat owners indeed observe market prices and stay ashore, especially in the summer season when a lot of smaller part-timers flood the market with raw materials when the weather is good. During this period it is common for full-timers to take out their boats for maintenance rather than fishing for lower prices. In turn, some fishers may take the risk and put out in rough seas, especially in areas where deep fjords provide shelter. Under normal conditions, these vessels would achieve lower prices, as buyers believe that fish caught in the fjords or close to the shore is of mediocre or poor quality (more on this issue in Chapter 8). When supply is short, however, the fish will be sold anyway, which creates a considerable economic incentive for fishers to take the risk. As one fisher puts it: 'You get few fish, but the price is almost double so that it covers the costs and your salary' (FN: 90). In this context, websites such as RSF.com provide information on average market prices that build 'prosthesis' (Çalişkan, 2007) for orientation and actual price realisation in the auction market (see Image 3.1).

At this point we are tempted to think that the economisation of the coastal fisheries through the organisation of new markets and property rights has truly liberated the once community-bound small boat fisher from her rural ties. While describing this transformation in the light of overfishing, neoliberal reform, social movements and local

Meðalverð

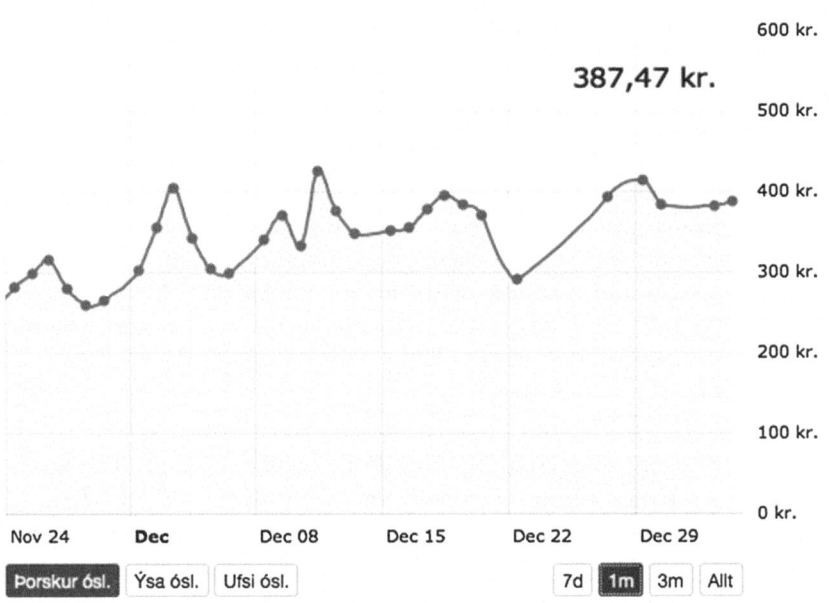

Image 3.1 Observing the auction market. On rfs.is, stakeholders can observe the average market price on the fish auction, in this case for cod (Þorskur). The image displays nicely the ups and downs in December 2014, in which several storms made fishing on small boats around the island state all but impossible for several periods. During this time, the market price usually increases due to lack of supply. With the beginning of the holiday season by the end of December, in which most boats don't fish, market prices typically sky-rocket and fall sharply as soon as the sea starts to calm down. (rfs.is, 5 January 2015)

practices that are manifested in the architecture of the small boat economy, this perspective does not tell is much about how the emergence of market institutions has transformed the economic practices of the small boat economy as such. Economising small boats, however, does not only transform access and property relations to natural resources, but also the socio-technical arrangements and practices that have broadened the horizon of possibilities beyond the geographical boundaries of their homeport communities, both at sea and ashore. In other words, the successively disentangling of the community-bound small boat fisher has created—to borrow and expression by Marx—a new class of 'double-free' small boat fishers who are free to buy, sell, lease and rent the exclusive right to fish; and free to sell their catch to whoever makes the highest bid. That this new form of rural independence is only to be achieved at the cost of a new form of monetised dependence will be demonstrated in the following sections.

Re-entangling Small Boats

With the successive integration of the small-boat fisheries into the ITQ system and the construction of fish auctions, small boat fishers have been disentangled and framed as fully-fledged market actor who can buy, sell, rent and invest in transferable fishing quotas and decide when to put to sea with regard to changing market prices. From this perspective, fishers are not naturally born as homo oeconomicus, but have been framed to behave successively *as if* they were rational economic actors in the neoclassical sense. We will now see, however, how this economisation of the small boat economy has not reconfigured the horizon of economic opportunities, but also the expectations on small-boat fishers who must adjust their practices to the new regime of market-based resource management in pursuit of monetary independence. As a consequence, the community-bound small-boat owner has not only become disentangled from his rural ties, but also *re-entangled* in a new techno-scientific network of monetised expectations and debt that henceforth condition the degrees of rural independence. As a consequence of playing along with the quota system, many small-boat owners such as Bjartur and his family who used to put to the streets to demonstrate the quota system have now become quota owners themselves and the tragic advocates of their consolidation.

Investing in Independence

Equipped with fishing quotas and the auction market, it seemed as Bjartur had finally become a truly independent fisher who could freely coordinate his fishing operations with regard to changing market prices. And even better, access was limited and quota shares allotted so he did not have to worry any longer about rushing out to sea every morning. Soon, however, it would become clear that it would be difficult to make a living from the allotted quota share, as the government would put on new regulations that closed down other fisheries that supplemented the cod quotas, which were increasingly reduced by the government on the advice of the Marine Research Institute. For Bjartur as for many other independent fishers this simply meant that fewer and fewer fish could be landed based on their share of the quota cake, as Bjartur recalls:

> And in the end [2004] they say: 'All boats who have days, now you have quota'. And that was not so much! And then fishers started to sell their quotas [while others bought quotas] and people start to make bigger plastic boats like they have and in Bolungarvík. And when we built [our new plastic boat], we bought quota, it was just in cod. But we could fish free in catfish and haddock, you could fish as much as you want in these kinds, and we were fishing 100 tonnes of haddock and 100 tonnes of catfish [Atlantic wolffish.[9]] In the community, we had 12 boats this time, but then the government decided to say 'Now we put in quota for haddock and catfish', and we have so short experience in catfish, we didn't get any quota...

It becomes clear that it is not merely 'the laws of the market' behind the decisions independent fishers take when confronted with a market-based management system. Rather, Bjartur had become faster re-entangled and constrained by a new web of markets, scientific stock predictions and political decisions than he could ever have dreamed of. This process of re-entanglement was the main reason why many fishers who only received small quota shares with the implementation of the ITQ system would sell out of the industry, as they did not see a way to make a living on only a few fishing days a year and rather cashed in on the sky-rocketing quota prices as starting grants for another life. This development was typical, especially for individuals with very low quota

---

[9]In Iceland, the Atlantic wolffish (Anarhichas lupus) is often referred to as 'catfish' in English.

shares from remote communities who did not have the means to invest in more quotas that would allow them to maintain a profitable year-round fishing operation.

Although Bjartur's family was left short with the implementation of the quota system, the dream of independence prevailed and the family would start playing along with the quota system by investing in more and more fishing quota and exchange their old wooden boats for a new state-of-the-art plastic vessel equipped with the latest technology. Those depending on jobs around the family business in the baiting house, the harbour or the local processing plant were grateful that the family would invest in the village. This decision, however, was not reason enough to convince other fishers to invest into the emblematic 'life in the harbour' that Bjartur idealises so passionately. As a consequence, Bjartur and his family were left alone with investing in their dream of rural independence:

> We had no quota for catfish, and we have like 16 or 20 tonnes in had-dock, and we buy 50 tonnes of quota for catfish, and the price, we got it on a very good price, the price just went up in 1–2 years. But our com-munity was then dead, because we had to stop fishing free in haddock and catfish. So the others stopped going out because they had no quota, so at one time I ended up alone here […], one boat…(MII)

The tragedy of Bjartur's home community is not an individual case of a single community, as many coastal fishers decided to sell off their quota share and leave to the capital region, contributing to an over-all decline of many remote communities around the country.[10] In a few other communities, however, some small boat owners rather saw the quota system as an opportunity for rebuilding their own com-munity on small boats. Especially fishers with better previous track records and quota share put them in a better position compared with others. Moreover, technological developments in fishing vessels, the

---

[10]For the impact of the quota system on regional development, see Benediktsson and Karlsdóttir (2011) and Dobeson (2019, forthcoming).

auction market and rising quota prices pointed towards a promising and profitable future and triggered a 'small boat revival' in some places typically equipped with good infrastructure and close proximity to the fishing grounds. But where did all the money for this unexpected revival come from?

Capitalising on Small Boats

Investing in independence is an expensive endeavour and would not have been possible without the help of local banks and the development of modern finance that allowed local fishers to stay afloat and invest in ever more expensive fishing quotas, new fishing vessels and state-of-the-art equipment. These loans played a crucial role for the small-boat revival in some coastal communities, in which local community-held saving banks ('Sparisjóður') saw the potential of investing in fishing quotas:

> So the bank, especially the bank here in our village was ready to lend money to those that they thought could make it as fishermen, and so they, the fishermen had to borrow a lot of money, but it was because the bank believed that they could pay back [...]. So they saw here, maybe the ones that were on the trawlers and they knew they were seamen and they knew they had what it took to be fishermen and they said to those guys: 'You invest with your money, I trust that you [pay back], I'll lend you some money, but also have to invest with your money to make sure that they would stay on the boat and do whatever they could to make it. (XXII)

While these types of investments allowed fishers to stay afloat and contributed to rebuilding some formerly deprived communities based on small boats, the translation of fishing rights into financial assets opened up a new window of opportunity for other actors from the capital region seeking for new sources of income and wealth. Especially in the midst of the Icelandic banking *útrás* (that is, the aggressive *expansion* of the Icelandic financial sector by the turn of the millennium) banks were more than eager to issue loans to almost anyone who could rely on some sort of income—and even better if borrowers had something relatively stable in value that could be used as collateral in case of default. As the wife of a small boat owner describes the atmosphere of the post-crisis era:

> In the gold rush before the crisis it was considered rather ungraceful to still follow the 'dirty' profession of the fisherman instead of increasing one's wealth with dubious banking transactions. Many sold off their quota shares and blew their money on sports cars. On the other hand there were those who tried pushing fishermen to borrow money to buy a lot of quotas. (MXI)

This development went hand in hand with the general development and role of the ITQ system, which spearheaded the wave of decentralisation and venture capitalism that hit Iceland in the mid-1990s, when the doctrine of financial deregulation became more and more influential in other sectors of the hitherto still very centralised economy. Thus, the integration of the small boat fleet into the quota system allowed banks to find cunning ways of capitalising on the rural areas by turning coastal fishers into investors with fishing quotas and vessels as valuable mortgages in their portfolios. As long as fishing quotas were kept scarce and there were still fish in the sea, one could bet on an increasing valorisation of fishing rights. In other words, highly valuable tradable fishing rights became a welcome asset as collateral in the accounts of the rapidly expanding Icelandic banking system. With a good fishing record from past years and a solid quota share, banks soon realised that even fishers with rather small boat quota shares were the perfect customers that would allow banks to capitalise on the countryside in the early 2000s, as Bjartur's mother Krístin recalls:

> BM: They came here in the year 2006 or 2007 from two banks, knocking on the door and say: do you want to change to our bank? We will give you more loan if you have security in your boat and my quota, and then you have money from us and then you can buy - what do you call it hlutabréf (stock) - that means paper in the bank.

It soon became clear that investments were not merely 'rational' decisions that mirror the activities of the quota market, as rural fishers were actively encouraged by the banks looking for valuable assets in their portfolios. Although Krístin claims to have declined the offer, it was these type of deals that allowed many fishers to invest in their independence as independent small boat fishers. At the same time, it gave fishers the liquidity to invest in new, ever fast and ever bigger small boats that would redefine what it means to be an independent small boat fisher, as another fisher hints at:

> So slowly, they [fishers] went from small boats to a little bigger boats, a little bigger, dealing with the quota system - taking it on as you could say. Because we here in the Westfjords, we were always very much against the quota system in the beginning, everyone was against it. (XXII)

'Taking it on', as quotation makes clear, however, not only involves investments in quota and gear, but may also involve playing the quota market and taking risks like a stockbroker or gambler at the casino, we will see in the following section.

Betting on the Future

The financialisation of fishing quotas has not only re-entangled small boat fishers into a new globalised web of money-mediated relations, but also redefined the cultural boundaries of what it means to be a small boat fisher who is now able to bet on the future and accumulate capital like any other capitalist investor. This new attitude is well reflected in the words of the CEO of one of the largest small boat owners known for aggressively investing and sucking up quotas from nearby communities:

> And it happens all over the world, you buy an apartment in Manhattan and the prices go up and you sell off, make a profit, so this is the same thing that happened with the quota system, but most, or many of the bigger firms today, they only bought and bought and bought quota, they didn't sell off, they just believed that it would be wise to buy up at those prices.

While the CEO legitimises the economisation of small boats to objects of investment and profit-making simply with the neoliberal zeitgeist in which market valuations have become a ubiquitous feature of social life, the quotation points at another crucial factor for understanding the unlikely revival of small boats in some coastal communities: the role of finance.

While doing fieldwork in one of the villages renowned for rebuilding the local economy on small boats, I asked a skipper how the recovery of the village, which was left devastated when a bigger trawler company sold out in the 1990s, had taken place. He explained that small boats had always been an integral part of community life that provided extra income and an important safety net that helped to buffer the devastating effects that hit the community when the large fishing trawler was moved. While small boats maintained a lifeline for the community, the community however was struggling until small boats themselves had become objects of high-risk business practices commonly associated with the elusive world of high finance. Accordingly, his cousin's family got lucky with playing the quota system with the help of the local banks during the expansion by speculating on market developments. Instead of simply investing more and more, however, he was clever enough to sell of his shares when the 'price was really high!' (XVI), just before the financial meltdown. As luck would have it, this enabled him to reinvest his money into even more quota shares to lower prices after the crash, which build the foundation for starting one of the largest small boat companies attached to a local fish factory in the village.

The example shows that fishing quotas have not only become a stable asset for banks, but also for the fishers themselves who are now able to place bets on the future in order to accumulate more capital to expand the boundaries of their freedom. Thus, the financialisaiton of fishing quotas has not only re-entangled small boat fishers into a new globalised web of money-mediated relations, but also redefined the cultural boundaries of what it means to be a small boat fisher who's independence is no longer solely depending on taking risks at sea, but also in the volatile world of modern finance: while speculation enabled few companies to grow excessively and rebuild their communities, at the same time it re-entangled fishers into a new form of globalised risk and volatiles markets similar to those of their large-scale counterparts. After a period of seemingly limitless growth provided by the mobilisation of lively capital from the sea, it was not until the meltdown of the banking system in 2008 that this new class of gambling small boat owners was confronted with the boundaries of their freedom. As both bigger and smaller companies started shaking due to the devaluation of the króna, financial instability was amplified by the fact that many of the loans that enabled the small boat revival were based on foreign currencies, especially in Japanese yen and Swiss francs. While some where lucky enough to have sold off their shares at sky-high prices just before the crash, others were not as lucky with playing the quota roulette. One of the latter is a skipper I will call Einar, with whom I established a closer relationship during my different stays in the field.

Einar was born and raised in the Westfjords and worked in every imaginable job in such an environment: he worked in processing plants, as an engineer and as a sailor on large ships. Like many other sailors, however, his dream was to become an independent small boat fisher. He remarks that this choice 'is very hard to understand, but I always wanted to be alone'. Consequently he invested in a small boat and got allotted a small amount of quota with the implementation of the system. As banks were eager to lend as much money as possible, he saw an opportunity not only to be an independent fisher, but also to build up a bigger company that could contribute to reviving his homeport and bring jobs to the community. Driven by this 'imagined future' (Beckert, 2016) spurred by the alchemists of modern finance, he could soon convince his brother, a medical doctor from a nearby village, to join the company: from now on, Einar could concentrate on the fishing, while his brother was in charge of all the finances and paperwork. As everything was going well and Einar was enjoying his new life a independent small boat fisher, the brothers could not resist the opportunity to expand their business even further by investing in more quotas and a new state-of-the art fishing vessel that would allow them to fish off their quotas more effectively. Opportunities seemed endless and banks were eager to support their vision by lending more and more money to the growing company, which soon would be one of the biggest small boat quota-owners in the region, with a share of approximately 800 tonnes of cod. All of a sudden, the quota shares exceeded by far what Einar used to fish during a single season, and even with his new boat this would have become and impossible endeavour. With all the investments the company grew rich in debt, but the brothers did not worry as it also grew rich in assets as quota prices seemed to rise endlessly. Moreover, profits started coming in more easily as they could lease out excess quotas to other fishers and companies in need of quotas who would fish off their share against a fee. In a way, Einar had become a rent-seeker who was gambling on the future—unfortunately very shortsightedly. With the crash in 2008, the tides had suddenly turned over night. For the two man company debt soared to entirely new levels and their creditors could not believe that the company could ever get back on track under given circumstances. As the bubble had burst, their bank decided that under no circumstances they could hold the company up and were forced to sell off everything at a much lower price—tragically to one of the large quota-owners in the region that was simply deemed 'too big to fail'. If they just had invested more!

While Einar and his brother, like many others got their 'heads cut off' (MXX) while other, often vertically integrated companies got 'their loans fixed', it becomes clear that we cannot simply account Einar's 'greed' for his failure, as many members of the community do. Rather, investing in the quota market and gambling on the future had become a structural necessity for staying afloat. Others, however, simply did not want to take the risk, as they did not see how to pay off the loans while at the same time paying the bills for crew, bait and boat. These fishers gave up their independence by selling their small shares and boat, often to one of the growing local quota owners for whom they would eventually start fishing as contractors.

Bjartur's family felt discontent with either option and just tried to mobilise as much capital as possible in order to stay afloat. Sure, with regard to the empty harbour, investing in a lot of fishing quotas and new fishing vessels might have been their only chance to bring back 'life in the harbour'. But after thinking it through and doing the maths, the family opted against this gamble, as Krístin who has strong opinions on what it means to run a family-owned company recalls:

We just stopped, we said: This is too high! - and we did not buy more… Because that was money from another country … And I said: no! We are small, we want to be small, no risk, just leave me alone (…), I said: Goodbye! - I was very lucky! (MVIII)

'Keeping it small' therefore is not only a matter of the size of the boat, but of the balance sheets of a company, which needs to be controlled accordingly. This 'control' is not only based on accounting, but also shaped by cultural valuations of what it means to different people to be a small-boat fisher. As we have seen, however, these boundaries can be difficult to navigate for rural fishers with little understanding or experience of modern finance whatsoever. Nevertheless, already smaller investments from this era put companies at serious risk of bankruptcy. In contrast to Einar who had to sell off quota and boat, however, Krístin's resistance to the idea of limitless growth would safe the company for now, as their moderate loans became part of recovery programme, in which debts were partly written off and converted into Icelandic króna to facilitate repayment, as she proudly explains:

They changed it because I had to pay the loan, it was Swiss money, euro and Japanese [yen]. I had to take another loan to pay everything and they change it to an Icelandic loan, Icelandic króna … And I have it on paper, they cannot come and say: you have to pay more, because I say: 'Now I'm finished, you have everything you need, I have payed you everything, you will let me have this money and I will always pay you every month and leave me alone', and they have to do it! (MVIII)

Although the family 'got lucky' under the circumstances, they are, like many other fishers and quota-owners today, still heavily indebted and literally fishing for their creditors. Although prices for fishing quotas plummeted after the meltdown of the Icelandic economy, they remained in high demand as valuable assets with regard to the insatiable global demand for protein from the sea. After all, however, the illusive world of modern finance is dependent on real people producing real commodities that realise real value in exchange relations with real people. As a consequence, those under the auspices of their creditors had become under immense pressure to extract as much value as possible from the sea in order to cope with the immense accumulation of debt after the crisis (see Appendix, Fig. A.4).[11] As the wife of a skipper who invested a lot of money in a new boat, quota and fishing gear put it,

'You could say that a lot of the boats you see in the harbours are actually owned by the banks, and the fishermen are forced to fish even in bad weather, as they are up to their necks in debt'. (MXI)

These tales from the field make clear that within the financialised world of market-based fisheries, it can be a very thin line that separates the dream of independence from the nightmare of bankruptcy. Although the investment opportunities provided by some local banks have helped in revitalising otherwise devastated fishing communities around a centuries-old and labour-intensive tradition, fishers and fishing communities have been increasingly tied to the volatile and inherently unstable world of globalised financial markets. The consequences for the small-boat economy will be sketched in what follows.

Economising Expectations

The re-entanglement of small boats into the globalised world of modern finance has completed the metamorphosis of small boats from *symbol and means of rural independence* to highly valuable *object of investments and profit-making*—and with it the cognitive frames and expectations that have transformed Bjartur and the other small boat fishers into entrepreneurs and investors. As a consequence of their financial entanglements in the form of debts and long-term liabilities, profit-making is no longer merely a welcome side-effect of rural independence, but a 'must-expectation'[12] to be fulfilled by those wanting to stay afloat.

---

[11]Although no differentiated data for the small-boat system are available, field interviews suggest a similar trend of investments and mergers.

[12]According to Schimank (2008), the increasing dependence of non-economic sub-systems such as politics, religion and education on money creates a pressure to economise their operations, which is characterised by changing expectations aligned with the goals of sub-systemic

Inevitably, quota owners and fishers have to find ways to economise their fishing operations in order to squeeze as much surplus value out of their allotted quota shares to serve their creditors.

Depending on the size and financial liquidity of the company running the vessel, quota owners can chose between two basic strategies to increase their revenues: investing new equipment that allows a more efficient extraction of resources from the sea; or reducing overhead costs, in particular for bait and wages. These two strategies cannot be viewed in isolation from each other, as the investment strategies that allow for more 'efficient' fishing operations enlarge the gap between two, as some evidence from the industry makes clear:

In general, economisation has led to a general trend towards higher operating costs in the small-boat sector based on investments in equipment, increasing fuel costs, maintenance and office costs and insurance that range between 75.4 and 92% of total revenues in the operating accounts in the largest vessel category of small boats under 10 metres[13] from 2003 to 2012, leading to overall thin profit margins when looking at the general picture after depreciation (Þórðarson & Viðarsson, 2014).[14] At the same time, operating costs reveal a general downward

---

goal-orientations. Accordingly, these changing expectations can be analytically distinguished in five different stages (ibid.: 13, translation by author): *Stage 1.* No consciousness of cost-orientation for system-specific performances. Actors with access to sufficient resources do not frame monetary liquidity as a problem; *Stage 2.* Cost-awareness becomes a 'should-be expectation' that frames actors' decision-making. Costs should be acknowledged and reduced in easy cases. *Stage 3.* Cost-awareness becomes a 'must-expectation' and activities must not produce economic losses. *Stage 4.* Profit-making becomes a 'should-be expectation'. Besides avoidance of losses, modest profit-making is welcome. *Stage 5.* Profit-making becomes a 'must-expectation'. Activities ought to produce as much profit as possible. Actors in public institutions typically follow the expectation-structure of stage 1 or 2. In the light of the economisation of the small boat economy, economisation of operation is to be understood as re-orientation of societal relations and future expectations of action towards profit-making.

[13]With close to 1600 registered vessels, the vessel category under 10 metres is by far the largest category of vessels equal or under 15 metres in 2013, followed by about 280 vessels between >10>13 metres and around 100 vessels <13<15 (Þórðarson & Viðarsson, 2014: 7; data include vessels without fishing licences).

[14]Accordingly, profits remained negative from 2003 to 2008 (except 2005), ranging from −1.8% in 2007 to a record loss of −82% in the crisis year (Þórðarson & Viðarsson, 2014). Interestingly, the trend seemed to have turned towards positive profits after 2008, from 5.1% in 2009 to 10.4% in 2010 and 4.9% in 2011, although 2012 suggests yet again a turn to negative profits

trend of wages for vessels under 10 metres from 42.9% in 2003 to 29.3% in 2012 to compensate for other increasing costs, in particular gear and oil as consequences of investments in new vessels, gear and changing market prices (ibid.).[15] Thus, while intensification enables the more effective extraction of resources, increasing investments and maintenance costs reinforce the pressure to economise, which is delegated down to the beginning of the value chain, in particular to fishers, line baiters and fish workers who must also find ways using resources more effectively—i.e. by working faster, taking risks at sea and adjusting their practices to the fluctuations of the market (more on this later).

The increasing pressure to economise operations has not only intensified the pressure on fishers, but also transformed the material culture of the small boat fleet, in which the traditional old wooden boats have been widely replaced by ever larger capital intensive high-speed fishing vessels (see Image 3.2).

This development is mainly driven by increasing capitalisation and concentration of fishing rights in the accounts of few large producers dominating the industry. Thus, despite the legal framework that intends to protect the small-scale fleet from their large scale counterparts, capitalisation and mergers from within have lead to similar developments that have radically altered ownership-structures and practices within the sector. Accordingly, data provided by the ministry of fisheries reveals a strong concentration of ownership rights in the artisanal H&L quota, where the ten largest operators own about 35% of the total quota quota-share (Fiskistofa, 2012). The businesses of these owners have very

---

(−3.8%) (ibid.). Although the data used above exclude the newer bigger small boats up to 15 metres that play a crucial role for the larger quota-owners in the small-boat system, it corresponds with interviews and field observations conducted during this study. Nevertheless, it is difficult to make general claims about profitability in the small-boat system due to the diverse history and background of fishing vessel owners, for example, that profitability is generally higher for those who got fishing quota allotted with the implementation of the system, whereas newcomers relying mostly or exclusively on costly quota-rentals are struggling. These figures, however, include all registered fishing vessels under 15 metres, thus including part-time vessels fishing in the summer coastal fisheries or lumpfish licences besides the market-based ITQ system (see below).

[15]The data indicate a rather steady increase in fuel costs from 2.4% in 2003 to 11.6% in 2012 (Þórðarson & Viðarsson, 2014). This can be explained by investments in newer vessel with stronger engines and increasing oil prices.

Image 3.2 'Small' longliner. The long liners Steinunn HF in the harbour of Flateyri, Westfjords, June 2014 (Photo by AD)

little to do with the traditional world of independent small boat fisheries, as the quota shares by far exceed what can be fished by an individual vessel during a season. Hence, after a peak in 2004 when all small boats were integrated into the ITQ system, the number of vessels in the ITQ-system declined steadily after a period smaller quota owners selling out and mergers within the small boat sectors that led to period of consolidation from 2010 (see Appendix, Fig. A.5). Typically, these companies represent relatively large vertically integrated companies owning a number of small-boats on which a number of hired fishers provide the raw materials for the local processing plant. Finally, the legal size of small boats in the H&L quota was doubled from 15 to 30 gross tonnes in 2014 under political pressure from large quota owners running processing plants in the small boat system, literally redefining the boundaries of what it means to be 'small'.

All in all, the economisation of the small boat economy has re-entangled all small boat fishers into a new globalised web of money-mediated relations that have made independence conditional on financial liquidity. While the financialisation of fishing quotas mobilised capital to revive some fishing communities with small boats, it at the same time transformed the meaning of coastal fisheries from symbol and means of rural independence to objects of profit-making and speculation. As a consequence, the immense accumulation of debt and long term liabilities pushes quota owners and workers in the industry to constantly economise their expectations about the future to stay afloat, thus reinforcing the cycle of investments, debt and intensification.

Reviving the Life in the Harbour?

The decision to play along with the quota system by investing in independence has not only alienated Bjartur's family from their former ideals, but also transformed themselves into belonging to a new class of petit capitalist quota owners entangled into volatile world of debt and global capital flows. And as this was not enough, especially townspeople embodied by the left-green avant-garde from the capital region started looking down on them, as Krístin laments: 'If I say I own quota, they just stop talking to me!' In Krístin's eyes, however, their ownership of fishing rights is legitimate as they 'bought the quota'—at least partly. Having taken larger loans to finance their independence, however, they have, as many other companies, developed a vested interest in maintaining the status quo while at the same time being opposed to reforms such as the long publicly demanded 'resource-rent tax'. In Krístin's eyes, increasing the costs of running their company by imposing extra financial burdens would make it even more difficult for smaller companies to stay afloat. Moreover, redistributing some fishing quotas to newcomers would mean to reduce the total amount of fishing quotas, likewise making life more difficult for keeping up with the payments to their creditors. Uncertain of what the future holds, any potential reform of the quota system appears as a threat to their existence, leaving them to be bitter defenders of the status quo. It seems as Bjartur has become the tragic hero in his pursuit of rural independence, knowing that defending the system undermines his dream of revitalising the harbour.

And indeed, increasing entrance costs due to sky-high quota prices and increasing operating costs have made it nothing but impossible to enter the industry for newcomers without access to vast sums of capital. In the eyes of many outside of the fishing industry, this de facto closure of the fishing is not only violation of the 'right to fish', but also the main material hindrance of rural development in many fishing communities. No matter what perspective Bjartur takes on the quota system, the prospect seems to be bleak. Nevertheless, some attempts have been made by the post-crisis government to accommodate the ongoing public criticism of the closure of the marine commons by reopening the fisheries to part-timers in the so-called Coastal Fisheries (Strandveiðar, or CF) over the summer months in 2009 (Halldórsson, 2010). Accordingly, any boat complying with the regulations for a professional licence can apply to fish in the system—also those allotted with ITQs. Registered vessels can land a defined maximum of fish per day until the total allotted quota for each fishing zone is finished and the fishery is closed.[16] While this attempt to reopen the fisheries has certainly filled up many harbour with small boats, the contribution of these boats to the local economy is relatively limited, depending on the number of boats in each zone, a few days every month from June to August. Moreover, many quota-owners as well as wealthy individuals in the ITQ-system have been increasingly entering the system as an additional source of income (Fiskistofa, 2015)—to the outrage of many locals and fishers. It is not uncommon for wealthy individuals from the capital region and former quota owners who sold out of the quota system to purchase a small vessel and hire a skipper to fish off the allotted quota. In fact, the number of regular quota-owners in the CF has increased steadily, with the exception of 2011, from 261 in 2009 to 362 in 2014. In contrast, the number of vessels in the CF without any quota steadily declined from 334 in 2010 to 146 in 2014 (ibid.), suggesting a low economic incentive for newcomers to enter the industry

---

[16]The total quota share for the CF system is split up and distributed among four different fishing zones for each season, starting from 1 May to 31 August. Each licence is only valid for one zone, which has to be decided before the season. Boats with a CF licence are only allowed to use handlines (see Appendix), that is, a maximum of four jigging computers for each boat. Boats must return to port after 14 hours, and their catch may not exceed 773 kg of cod.

and operating vessels exclusively for the summer fisheries. It is therefore questionable to what extent this attempt to partially reopen the fisheries has really contributed to reviving the iconic *life in the harbour*.

Conclusion: Unfulfilled Expectations

This chapter has shown that the economisation of the small-boat fisheries must be understood in the broader historical context of a culture of small-boat fisheries, which were successively economised and transformed by means of market-based ideas on modern resource management. It has been shown that the organisation of such markets is the result of political power struggles over the right to fish and presupposes the disentanglement and translation of rural fishers from their rural ties into fully-fledged market actors who can freely engage in market transactions. On the other hand, the chapter has shown how this economisation process is prone to stimulate a prompt re-entanglement of fishers, who engage in new investments to expand the boundaries of their independence in the new regime of liberal rural capitalism. This, in combination with the deregulation and financialisation of the Icelandic economy, has stimulated massive investments in fishing gear and quotas, while at the same time opening up to speculations on rising prices for fishing rights. With the meltdown of the Icelandic economy in 2008, however, the quota bubble burst and many companies who engaged in speculation on rising quota prices were doomed to fail. On the other hand, other quota-owners who either sold out of the industry before the crash or only had moderate debt profited from post-crisis prices and could remain afloat. These companies, however, are tied to their financial liabilities and are literally forced to fish off their debts and squeeze as much surplus value as possible out of their quota shares to stay afloat. Thus, the culture of independent small owners is being challenged by tendencies towards increasing capitalisation and rationalisation, which has changed the economic expectations from general cost-awareness to an imperative towards profit-making in an increasingly professionalised year-round small-boat fishing industry that has put small-boat fishers at the mercy of global market movements and volatile capital flows.

It has furthermore become clear that the economisation of the small-boat fisheries into valuable assets not only implies a re-entanglement in a money-mediated system of profit-oriented expectations, but also a transformation of the culture of small-boat fisheries that has turned the former opponents of the quota system into investors and tragic defenders of it. While small boats still signify the dream of rural independence for fishers and quota-owners at the rhetorical level, the reality of today's small-boat fisheries looks very different, as the stories of Bjartur and the other remaining fishers have shown. In a way, one could say that they are now forced to be independent—as agents of economisation and profit-making.

All in all, the culture of small-boat fisheries is increasingly losing its role as symbol of independence and significance as a safety net for the coastal communities in decline. This development can also be observed on the political level, where new lobby groups of larger quota-owners have started challenging the privileges and protective measures of the small-boat system, which implies a fragmentation of social solidarity that once led the political struggle to protect the culture of small boats. The consequences of this development can in particular be observed on the political level. At the time of writing, a new lobby group of larger quota owners was about to split from NASBO, as they no longer wanted to follow the one-vote-per-boat rule within the organisation. Furthermore, a new lobby group (Landssamband línubáta) of larger quota owners using baiting machines have split from NASBO to challenge the quota discount for hand-baited lines—a decision that would most likely put an end to hand baiting, putting communities in decline at risk to lose even more jobs due to rationalisation.

Another sign of this cultural transformation is that it is almost impossible for newcomers to enter the industry. Although the market system is in principle open to everybody, entrance costs have become way too high for anyone without fishing quotas or other financial assets.[17] This fact has caused a lot of frustration among the increasing

---

[17]According to estimations from NASBO, the minimum entrance cost for entering the small-boat quota system in a profitable way lies around 190 million ISK for a boat with quota—about €1.273 million at the time of writing (Stefánsson, 2015).

workforce of fishers who need to sell their labour on a contract basis to one of the quota-owners. However, the desire to be one's own master remains strong, especially for those contracted to one of the larger companies. All in all, the situation remains difficult for those left out. Although buying a small boat and renting costly quota seems to be an option for some, investing in quota at the current market prices seems all but impossible. As a successful and respected small-boat skipper puts it desperately with regard to his chances of being becoming his own master:

> Why can't the guys that go to sea own the quota? If we would go to a bank and ask for a loan, they would laugh at us. (FN: 75)

Today, the only option for entering the industry is through the non-market system, in particular the part-time summer coastal fisheries. Some of those who invested in a small boat for this system did so because they speculated that these boats would eventually be integrated into the ITQ system. Lobbying efforts from quota-owners, however, are strictly opposed to this, as this would mean cutbacks on their own quota-accounts, and, as it stands today, there seems no sign that the hopes of these coastal fishers will be fulfilled. Others saw the summer coastal fisheries as a chance to finance a fishing vessel as a starting point to slowly get a foot into the industry, as a young fisher from a thriving community who invested in a small boat without quota together with his father: 'I am trying to own something, maybe later I get, when I finish pay the loan', (MXVI). Because of low quotas and only a few days at sea in the summer coastal fisheries, however, competition for the limited quota is fierce[18] and every missed open day at sea means a significant loss for him who has to work another job to keep up the payments for the loan. For this reason, fishers who betted on the summer coastal fisheries and took loans to get a foot into the industry are often under immense pressure to secure their free quota share in order to pay off

---

[18]This is especially true for vessels stationed in the Westfjords region, which is the fishing zone with the most vessels.

their liabilities. In this system, however, fishers cannot choose when to put to sea and therefore often do so in bad weather, turning the fishery into an often brutal grind. As one skipper, who was almost dozing off after spending four long days at sea told me: 'I am getting really tired of it' (MIX), indicating that his hopes and expectations for the part-time system had vanished.

While this chapter has shown how the economisation of the small boat economy has transformed small boats from symbol and means of rural independence to objects of investments and profit-making, it still remains to be seen how our new cultural figure of the independent small-boat owner is coping with these new entanglements in practice. The next chapters will therefore turn to the daily life at shore and at sea to investigate how the profound cultural transformation of the world of modern small-boat fishing is gaining momentum.

References

Ahrne, G., Aspers, P., & Brunsson, N. (2015). The Organization of Markets. *Organization Studies, 36*(1), 7–27.

Apostle, R., Barret, G., Holm, P., Jentoft, S., Mazany, L., McCay, B., & Mikaelsen, K. (1998). *Community, State and Market on the North Atlantic Rim: Challenges to Modernity in the Fisheries.* Toronto: University of Toronto Press.

Arnason, R. (2008). Iceland's ITQ System Creates New Wealth. *The Electronic Journal of Sustainable Development, 1*(2), 35–41.

Aspers, P. (2009). *How Are Markets Made?* Retrieved from Köln.

Beckert, J. (2016). *Imagined Futures: Fictional Expectations and Capitalist Dynamics.* Harvard: Harvard University Press.

Benediktsson, K., & Karlsdóttir, A. (2011). Iceland: Crisis and Regional Development—Thanks for All the Fish? *European Urban and Regional Studies, 18*(2), 228–235.

Çalışkan, K. (2007). Price as Market Device. Cotton Trading in Izmir Mercantile Exchange. In M. Callon, Y. Millo, & F. Muniesa (Eds.), *Market Devices* (pp. 241–260). Oxford: Blackwell.

Çalışkan, K., & Callon, M. (2009). Economization, Part 1: Shifting Attention from the Economy Towards Processes of Economization. *Economy and Society, 38*(3), 369–398.

Çalişkan, K., & Callon, M. (2010). Economization, Part 2: A Research Programme for the Study of Markets. *Economy and Society, 39*(1), 1–32.

Callon, M. (1986). Some Elements of a Sociology of Translation: Domestication of the Scallops and the Fishermen of St. Brieuc Bay. In J. Law (Ed.), *Power, Action and Belief: A New Sociology of Knowledge?* (pp. 196–223). London: Routledge.

Callon, M. (1998). *The Laws of the Market*. Oxford: Blackwell.

Callon, M. (1999). Actor-Network Theory—The Market Test. *The Sociological Review, 47*(1), 181–195.

Callon, M., Millo, Y., & Muniesa, F. (Eds.). (2007). *Market Devices*. Malden: Blackwell.

Dobeson, A. (2019, forthcoming). Das Fischerdorf im liberalen Kapitalismus: sozialräumliche Öffnungs- und Schließungsprozesse in der nordatlantischen Peripherie. In A. Steinführer, L. Laschewski, T. Mölders, & R. Siebert (Eds.), *Das Dorf. Soziale Prozesse und räumliche Arrangements*. Berlin: LIT.

Einarsson, N. (2011). Fisheries Governance and Social Discourse in Post-crisis Iceland: Responses to the UN Human Rights Committee's View in Case 1306/2004. *Yearbook of Polar Law, 3*, 470–515.

Eythórsson, E. (2000). A Decade of ITQ-Management in Icelandic Fisheries: Consolidation Without Consensus. *Marine Policy, 24*, 483–492.

Fiskistofa. (2006). *The Fisheries Management Act*. Reykjavík: Icelandic Ministry of Fisheries and Agriculture. Retrieved from http://www.fisheries.is/management/fisheries-management/the-fisheries-management-act/.

Fiskistofa. (2012). *Aflahlutdeild stærstu útgerðanna*. Retrieved from http://www.fiskistofa.is/umfiskistofu/frettir/nr/775.

Fiskistofa. (2015). *Yfirlit yfir strandveiðar 2014*. Retrieved from http://www.fiskistofa.is/veidar/aflaupplysingar/yfirlit-sidasta-fiskveidiars/strandveidar/.

Fligstein, N. (1996). Markets as Politics: A Political-Cultural Approach to Market Institutions. *American Sociological Review, 61*(4), 656–673.

Fligstein, N. (2002). *The Architecture of Markets*. Princeton: Princeton University Press.

Graham, I. (1998). The Emergence of Linked Fish Markets in Europe. *Electronic Markets, 8*(2), 29–33.

Halldórsson, G. H. (2010). *Strandveiðarnar 2009: Markmið, framgangur og fiskveiðistjórnun* (Master's thesis). University of Akureyri, University Centre of the Westfjords, Ísafjörður. Retrieved from https://skemman.is/handle/1946/5668.

Helgason, A., & Pálsson, G. (1997). Contested Commodities: The Moral Landscape of Modernist Regimes. *The Journal of the Royal Anthropological Institute, 3*(3), 451–471.

Ingimundarson. (2008). Fighting the Cod Wars in the Cold War: Iceland's Challenge to the Western Alliance in the 1970s. *The RUSI Journal, 148*(3), 88–94.

MacKenzie, D. (2009). Constructing Emission Markets. In D. MacKenzie (Ed.), *Material Markets: How Economic Agents Are Constructed* (pp. 137–176). Oxford: Oxford University Press.

Schimank, U. (2008). Kapitalistische Gesellschaft - differenzierungstheoretisch konzipiert. Beitrag zur Tagung der Sektion Wirtschaftssoziologie der Deutschen Gesellschaft für Soziologie "Theoretische Ansätze der Wirtschaftssoziologie", Februar 2008, Berlin.

Statistics Iceland. (2011). The Fishing Fleet at the End of 2010. Retrieved July 20, 2015, from Statistics Iceland http://www.statice.is/lisalib/getfile.aspx?ItemID=12164.

Stefánsson, K. (2015, February 5). Hvað kostar að gerast trillukarl? *Fiskifréttir.*

Sverisson, Á. (2002). Small Boats and Large Ships: Social Continuity and Technical Change in the Icelandic Fisheries, 1800–1960. *Technology and Culture, 43*(2), 227–253.

Þór, J. Þ. (2002). *Sjósókn og Sjávarfang. Saga Sjávarútvegs Á Íslandi.* Akureyri: Bókaútgáfan Hólar.

Þórðarson, G., & Viðarsson, J. R. (2014). *Coastal Fisheries in Iceland.* Retrieved from http://www.matis.is/media/matis/utgafa/12-14Coastal-fisheries-in-Iceland.pdf.

4

The Practice of Fishing

It's not rocket science!

Deckhand

On a bright polar day in June, I arrive at the harbour around midnight, where a local skipper is waiting for me to join him on a longlining trip. In contrast to Bjartur, the skipper, like many others putting to sea today, does not own the means of production and is hired to fish off what is left of the owner's quota for the season. We grab our provisions for the trip, board the boat and drive over to the other side of the harbour, where the other member of the crew, the deckhand, is already awaiting us at the docks to put the 32 *bala*—buckets of longlines—on board with the help of a small crane at the docks, making a total of 9 kilometres of longline and 16,000 hooks that were hand-baited in the village. After finishing his job, the deckhand boards the 14.96 tonne and 12.45 m long boat and we are ready to sail.

© The Author(s) 2019, corrected publication 2020
A. Dobeson, *Revaluing Coastal Fisheries*,
https://doi.org/10.1007/978-3-030-05087-0_4

After gathering at the wheelhouse, where the skipper will make the obligatory call to register that his vessel has put to sea at the Icelandic marine administration, the skipper will inform the crew that he has decided to sail much further than usual, about 60 miles offshore, just above the Arctic Circle, as the forecast is quite good and 'heavy lines' were reported from that area recently, which used to be reachable only for the large steel vessels of the fleet just a few years ago. It will turn out during the trip, however, that it is not only the range of the small-boat fleet that has changed over the years.

Although the technique of longline fishing has essentially remained the same for centuries, the modern world of small-boat fisheries has little to do with the world of open rowing boats of pre-modern days, when fishers had to rely on nothing but their faith in God and experience of 'reading nature' before putting their lives and fortunes at the mercy of the sea. In this—to borrow an expression of (Foucault, 1970/2002)—*order of things*, fishing was a duty before God, who, if merciful, maintained the livelihood of the people, rather than being a source of wealth and prosperity in the modern sense.[1] The fish were caught, gutted, salted and stored by the same men at the fishing station, where they would try to make a living during the winter months before they returned as peasants to their mainland farms in the spring.

Today, longline fishing on small boats is a highly professionalised year-round economic activity that takes place in a highly industrialised and technologised environment. By joining a fishing crew on their daily grind, this chapter makes a humble attempt to understand the world of modern small-boat fishing from an 'inside' perspective. The argument for doing so is simple and straightforward: actual fishing forms the backbone of the entire fishing industry. Hence, the value of the quota and the raw material is essentially grounded in the doings and landings of the fishers. Without people putting to sea, there would be

---

[1]A historical artefact of this ancient world around an ancient fishing station be found in the Westfjords at the Ósvör outdoor museum near Bolungarvík. As one of the oldest documented fishing stations in Iceland Ósvör was rebuilt for an excellent historical documentary entitled *Íslands þúsund ár* (1997; English title: *Give us this day*) that gives an insight into life and work around the station before the advent of capitalism and the motorisation of the fishing fleet.

no wharf building fishing vessels, no fish markets, no fish processors, no quota market, no global exchange relations and not demand. Put differently, our understanding of economisation and marketisation remains incomplete without understanding how it is materialised at the outset.

The ethnographic analysis presented in this chapter stands in light of what has been widely summed up under the label 'Theories of Practice' (Reckwitz, 2002; Schatzki, 1996; Schatzki, Knorr Cetina, & Savigny, 2001). Accordingly, social order can be explained neither as the sum of cognitive and conscious interaction patterns between atomised individuals, nor as some abstract reality *sui generis* that is separated from human agency. Thus, the world of fishing is not simply a more or less static, taken-for-granted, objective lifeworld separated from the mind (Schütz & Luckmann, 1975/2003), nor blindly reproduced as incorporated habitus (Bourdieu, 1977). Rather, we will see that the world of fishing is grounded and reproduced in the contingent situatedness of socio-technical practices, which allow fishers to skilfully cope with and adjust to an ever-changing, potentially dangerous and highly volatile work environment.

The Floating Workshop

Once the vessel leaves port, the crew and their vessel will remain tightly knit until they return. The physical boundaries of movement are clearly constrained by the material expanse of the vessel itself from the stern to the bow, from port to starboard. If the crew crosses these boundaries, there is nothing but the icy water and the strong currents of the North Atlantic, bearing the serious risk of hypothermia or drowning within minutes.[2]

While the novice fisher on board seems to be a little nervous about putting his life at risk in a plastic hull for the next 15 hours, the skipper and the deckhand do not seem particularly concerned about the deep

---

[2]To prevent hypothermia in case of emergency, all fishing vessels in Iceland are required by law to be equipped with special floating dry suits.

reflections of their new crew member. Instead of feeling separated from their material environment and thinking about all the potential perils of the sea, the mood of the crew is rather relaxed as they engage in fairly mundane activities, such as putting away the provisions, starting the engine, checking the weather forecast, talking on the wireless, discussing fishing strategy and exchanging the latest community gossip. In other words, the crew does not seem to be aware of the fishing vessel in the same sense as the ethnographer, who has already objectified the vessel and the crew in his mind. Instead of being merely a constraining space, the relation of the crew to their environment seems to be first and foremost of a practical nature and reminds one of the doings in a workshop—a floating workshop, in which tools and machinery are used for harvesting and manufacturing raw materials.

As in any workshop, there are different areas for work, such as the wheelhouse, the deck and the hold, which all have different functions in the manufacturing process. In contrast to the traditional conception of a 'stationary' workshop in which goods are crafted with the help of tools, however, the floating workshop goes beyond manufacturing as it provides shelter from the sea and allows for the spatial relocation of the workshop through navigation. But how can we understand this mundane relation of fishers with their workshop?

Skilful Coping in the Wheelhouse

As the name indicates, the wheelhouse is not only the central site for social gathering, but also for navigation, communication and control. Today, the architecture of the wheelhouse allows for a 360° view around the boat, while at the same time displaying a multitude of information via electronic devices providing information about course, weather forecast, fuel, level of sea water on board and the state of the engine room (see Image 4.1). Based on similar observations from modern marine navigation and the aviation industry (Hollan, Hutchins, & Kirsh, 2000; Hutchins, 1995a, 1995b), scholars have made the point that cognition is *distributed* among socio-technological arrangements.

Image 4.1   The wheelhouse as focal point of circumspection. The skipper monitors the wheelhouse while the boat is sailing on autopilot towards the fishing grounds. The skipper has full responsibility for the boat and always keeps an eye on the equipment above the steering wheel, although most devices have a voice alarm (for example, when too much sea water has flooded the hull). To the upper right, the monitor displays the engine room, which according to the skipper is 'very important. You really don't want a fire on your boat'. The computer screen to the upper left and the two to the lower left have multiple usages and programmes can be switched according to individual needs. To make his monotonous job more bearable and because the sea is fairly calm, the skipper has decided to watch a film (upper left) and to get in contact with his social network community (lower middle). This, however, does not imply that the skipper is losing awareness of his surroundings (Photo by AD)

Cognitivist theories of perception that include cultural-material aspects of cognition (Hollan et al., 2000; Hutchins, 1995a, 1995b; Knorr Cetina, 1989), however, by definition presuppose one or another form of representation in the minds of intentionally—in the technical sense of object-directed—subjects (Dreyfus, 1993, 2004). There is, however, no evidence that the skipper engages in some more detailed analysis or distributed 'computation' (Hutchins, 1995a) of information or fixation of a certain aspect of reality, as the skipper's orientation towards many tasks implies dispersion within a unified environment rather than fixation on an isolated aspect of reality, as the following example illustrates.

While leaving port, the skipper must manually navigate his vessel by utilising the steering wheel and the lever. To do so, he does not first have to understand the mechanical functioning of the devices. Hence, when the skipper is utilising the wheel and the lever, it seems that the lever somehow 'disappears' as an object of a theoretical nature (Knorr Cetina, 2001: 178). The observations, however, suggest that the relation of objects as representations and human practices is reversed, meaning that objects are not first conscious and then disappear in the sway of practical coping. Rather, the wheel and the lever seem to build a relation with the skipper that is prior to his theoretical understanding. We can therefore say with Heidegger (1962: §15) that the wheel and the lever—like the famous hammer in Heidegger's workshop—are essentially *ready-to-hand*. Hence, instead of being perceived as two separated objects, wheel and lever can be better understood in terms of what Heidegger (ibid.: §15: 97) calls *equipment*. Accordingly, equipment is defined formally as something 'in-order-to' (ibid.): the wheel is equipment *in order to* change the direction of the boat; the lever is equipment *in order to* change the speed of the boat. In this sense, the wheel and the lever are not isolated atoms in an objective space, but gain their meaning in relation to other equipment, such as the sonar, the propeller, the buoys, the fishing lines—in other words, in the *totality of equipment* (ibid.), which itself is equipment in order to catch fish: the floating workshop of the fishing vessel. Hence, instead of presupposing first engagement in some rule- and object-oriented behaviour, it rather seems that the relation of the skipper to the wheelhouse is rather grounded in a pre-intentional (what one might loosely call pragmatic) relation with equipment, which Heidegger (1962: §16, §69) has called the *circumspection of concern*.

The circumspection of concern designates a pre-reflexive form of involvement with the world on the part of human beings (Heidegger refers not to human beings but to *Dasein*—literally 'being-there'—in order to try to get to the phenomenon without being pre-empted by the familiar subject/object pattern). Pre-reflexivity, however, does not mean that circumspection remains 'sightless', as Heidegger (1962: §15: 99) remarks. Rather, circumspection designates a state of dispersed awareness within the surrounding environment. It is for this reason that the skipper can engage with a multitude of equipment at the same time: while steering the boat, the skipper will talk on the wireless to other fishers, check the control panel and retrieve information from his chart plotter and navigation software.

Thus, in contrast to the 'theoretical gaze', which Heidegger characterises as 'just looking' (ibid.: 98) at an isolated aspect of reality, the skipper feels with his body how to move the boat against the current by being constantly attached to the physical movement of the vessel, while at the same time being aware of the vessel's distance from the docks—just as an experienced driver has a sense of distance and space when parking his car without deliberately measuring the distance to the car in the next parking lot. Space, however, is not encountered in the Cartesian sense of a container-like extension, but as a qualitative phenomenon, which shows itself first and foremost as what Heidegger calls 'de-distancing'[3] or 'bringing-close' of something (Heidegger, 1962: §§22–24). In this sense, the vessel can rather be described as an extension of the skipper's body that allows him to gauge distances and movement in the form of a bringing-close, rather than being a cognitive attachment in an extensive spatial coordinate system of abstract relations.

In this sense, the observations from the wheelhouse confirm the deckhand's statement that the doings on board do not involve 'rocket science'. Nor does the skipper seem to engage in any form of theoretical reflection that separates him from the boat and the environment. It would be wrong, however, to assume that fishing is 'easy' and does not require any form of skill. In fact, it is physically demanding and requires considerable mastery in order to ensure a smooth coordination

---

[3]The original translation of *Sein und Zeit* renders this 'de-severence', but Hubert Dreyfus' translation of 'dis-stance' seems more accurate.

of activities around heavy machinery in a potentially rough and dangerous environment. Hence, experienced fishers know how and when to use the machinery 'by heart', just as they have a feeling for how to adjust their bodies to a rolling boat while working on deck. In line with Hubert L. Dreyfus' Heideggerian phenomenology of everydayness (Dreyfus, 1991, 1993; Dreyfus & Dreyfus, 1984), I will therefore refer to this specialised involvement with equipment as *skilful coping*.

In contrast to Bourdieu's (1977)[4] class-based notion of *habitus*, which reduces the logic of practices to incorporated forms of pre-conscious 'regulated improvisation' as the basis for the symbolic reproduction of a given social order, skilful coping refers to more general *mode of awareness* (ibid.) and *comportment* that is not to be confused with 'mindless, mechanical behaviour' (Dreyfus, 1993: 88–89). Using the example of the skipper, skilful coping therefore cannot be reduced to blindly following certain values, understandings, aesthetic tastes imposed by the 'social' logic of a field, but a socio-material practice that allows flexible adaption to the contingent and potentially dangerous environment in which fishing takes place.

Thus, rather than first needing to engage in some sort of mental act or calculation before moving the vessel, the observations from the wheelhouse suggest that the skipper is always already situated *in* an already meaningful disclosed world that lies before any form of subject/object divide (Heidegger, 1962: §§14–27). Thus, the skipper does not need to first understand his environment in the sense of a conscious subject in order to have a feeling for the movement of the boat, as the world is yet already disclosed and understood. Heidegger (1962: §§14–27) has called this primordial relation of man with the world *being-in-the-world*. This, however, does not merely imply a primacy of practice over theory, but points at a more fundamental relation of our skipper with the world, as the following section makes clear.

---

[4]Needless to say, Bourdieu's practice theory was strongly influenced by Heidegger's phenomenology of everydayness. In contrast to other practice theorists, such as Dreyfus and Schatzki, however, Bourdieu never really engaged in developing a fully-fledged theory of practice in the tradition of the early Heidegger and therefore seemed to return more and more to the French structuralist tradition he had once departed from.

The World of the Fisher

The further we go, the rougher the sea becomes and the boat starts rolling heavily. As an inexperienced novice, I make the mistake of leaving my lunch box unsecured on the table and only manage to save it at the last moment from being spread all over the wheelhouse when a large swell hits the boat and our bodies are forced out of their comfort positions. The skipper smiles at me and says: 'You cannot just watch a movie on autopilot in bad weather', indicating that his involvement with the computer screen has just been shifted to the sea state, which caught his attention by being mediated through the vessel to my lunch box and our bodies, in which our primary involvement with the world is grounded (Merleau-Ponty, 1945/2012).

This little episode makes clear that the skipper is not merely a mental subject that engages in conscious observation of his environment. Rather, it highlights a form of pre-reflexive involvement with the world, which is not necessarily based on the direct use of artefacts and material devices. In this sense, Heidegger (1962: §18: 118) remarks that 'involvement' 'is itself discovered only on the basis of the prior discovery of a totality of involvements' and it is this 'pre-discoveredness' in which 'there lurks an ontological relationship to the world' (ibid.). This implies that the world is always already pre-understood in a non-reflective way: the skipper already has an understanding of how the boat moves, in which weather conditions and what this means for his involvements and dealings at the wheelhouse. In this way, the world does not simply 'act' on the skipper in the sense of stimulus and response, as any stimulus from the world requires a pre-understanding in order to be perceived at all. In this sense, the perception of the factual world is itself based on a phenomenal structure, which (Heidegger, 1962: §14) refers as the *worldliness of the world*.[5] This implies that the world does not belong to some external objective sphere, but to the

---

[5]Here I follow Dreyfus' (1991) translation as it literally refers to the German term '*Weltlichkeit*' rather than using the term 'worldhood' as used in the English translation of *Being and Time* from 1962.

skipper himself. In other words, the skipper is not only *in* a factual world of socio-material relations. Rather, the analysis suggests that the skipper himself *has* a world.

'Having' a world implies that the relation to the world cannot be described in terms of mere behaviour, as seems to be the case with most animals. In this sense, a fish's behaviour takes account of the currents of the sea, but it remains bound by its physical environment. And although machines that make up the fishing vessel may have some sort of primitive actor quality, their functioning requires a construction plan (Heidegger, 1983: §51). The world of man (Dasein), however, lacks any construction plan, although in one aspect it is its constructor (ibid.: §42). In other words, having a world implies that humans have to constitute their own world, as they are not only bound to their physical environment, but able to reflect, understand and transcend the factual world.[6] It is this constitutive openness and self-reference that characterises our primary relation with the world. Thus, instead of simply assuming methodological symmetry between human- and non-human actors (Latour, 2005), the practice theoretical account deployed in this chapter is directed to the practices of humans *with* their socio-material and 'more-than-human world'.[7] Only from this 'human-centric' perspective can we understand how the world of fishing gains its momentum through daily coping with an ever-changing and increasingly technised and economised environment.

In the next section therefore we illustrate how the skipper encounters and deals with his factual world at sea, using the example of disturbances and malfunctioning of equipment, which can put the lives of the crew at serious risk and danger from one moment to another.

---

[6]Today, it seems undisputed that most animals are in some form consciously aware of their environment. There are however good reasons to assume that the ability to reflect and transcend the factual world is most developed in humans. This, 'eccentric' feature of man, which builds the foundation of metaphysical thinking and creativity that allows humans construct and deconstruct their own world does not per se imply an essentialising ontological difference or clear-cut boundaries between humans and non-humans (see Dobeson, 2018).

[7]We will explore the more-than-human entanglements of human coping in Chapter 6.

When the Fishing Line Gets Tangled

While the skipper is absorbed in a state of skilful coping, a squawking sea gull will not draw any special attention, although he might well hear it. The seagull bears no reference to the equipment ready-to-hand and does not interfere with his coping. In contrast, a screaming alarm from the wheelhouse—for example, when too much seawater has flooded the hold—will attract the attention of the skipper, as the signal bears a reference to the safety of the vessel. Usually, the signal will stop, as the seawater is pumped automatically out of the vessel and the skipper will just maintain his routine at the wheelhouse, as he is used to the alarm occurring when sailing through rougher waters. In case of a persisting alarm, however, the signal becomes *conspicuous* and shows itself in the mode of *obtrusiveness* (Heidegger, 1962: §16). Hence, the skipper's attention is shifted from what Dreyfus (1991: 70) refers to as 'absorbed coping' to 'deliberate attention' in case the hold actually is flooded, a fishing line is snagged or tangled (see Image 4.2), the steering does not work or the engine breaks down.

In these drastic cases, the reference of equipment to the totality in which it has its being or makes sense becomes obvious, as the lives and safety of the whole crew are at stake. For this reason, deliberate attention will be focused on restoring the primary function of equipment. In these cases, equipment becomes *present-at-hand* (*vorhanden*), that is, isolated and problematised by means of reflection and objectification (Heidegger, 1962: §16, §69b), which is also a basis for causal problem-solving. This, however, does not mean that the object is separated from the world as abstract entity; abstraction itself is only possible in the context of the world in which it is essentially grounded.

If the problem can be fixed, practical problem-solving will eventually fuse back into absorbed coping, for example, when the boat has been stabilised, the steering wheel has been fixed, the fire in the engine rooms is under control or the fishing line has been repaired. If the problem remains, however, the fishing line will probably be lost, or the crew must call on other vessels and the coast guard to either tow them back to port or rescue them. In any case, however, 'making sense'—practical

Image 4.2 When the fishing line gets tangled. After the first string is almost hauled in after more than three hours of non-stop work on deck, the routinised order is broken and turned into a tense state of danger as the fishing line and the anchor line of the buoy turn out to be entangled and stuck in the winch. The risk of losing the buoy and the fishing line can be a substantial economic loss to the crew, but it is also a threat to the safety of the crew on board the fishing vessel as line tension is growing on the winch. The skipper and the deckhand must now improvise and see whether the fishing line and the buoy can be safely disentangled. The skipper reacts rapidly by pausing the grid winch, freeing the lines by cutting them with the knife used for cutting the fish, tying the loose ends back together with the assistance of the deckhand securing the lines from not slipping away and putting the repaired end back on the winch. Shortly afterwards, the rest of the fishing line is brought back in and the order of routine and practice seems restored. While looking more than relieved, the skipper merely remarks 'You can imagine how it is if this happens in bad weather!' (FN: 78) (Photo by AD)

and reflexive—can never be seen merely as the result of some sort of isolated cognitive act in which the world as external entity 'reacts' on a closed cognitive system. Nor can we simply presuppose a primary relation of practice over theory, as both stances are fundamental features of human coping (Heidegger, 1962: 238).

In this section we have pointed out different forms of involvement on the part of the fishing crew, which are based on a set of routinised practices and experience at sea. But how does a novice acquire the skill to cope with this rough and unpredictable environment?

The Novice

Training a novice with little or no experience at sea is always a risk for a skipper as he does not know beforehand how his apprentice will cope with the new environment. Especially in rough seas, seasickness seems to be a problem for novices, who either quit the job or learn how to cope with it over time. In case of nausea or any other reason for malaise, the skipper has to show solicitude for the novice and make sure that he can rest. But what is the difference between skilful coping and novice coping?

We have already pointed out that the novice's perception of the fishing boat as a new place that needs to be discovered and understood is different from the 'absorbed' state of being and coping that characterises daily and routinised activities. The examples above have shown, however, that not all coping takes place in a state of absorbed coping, as irregularities may arouse more deliberate attention on the part of the skipper. But it is especially the novice on board who engages in many more reflexive acts such as observing the actions of the skipper and the deckhand to try to understand *how things are done*. Furthermore, he will try to follow general advice and orders from the crew and try to remember rules and guidelines and try to acquire skill by imitation and repetition.

In order to acquire the skill of longline fishing, however, the crew will show the novice by demonstration how to execute the tasks that structure the fishing operation. For the sake of safety, the novice will of course start with easier tasks such as gaffing (see Image 4.3) and cutting the fish, but it is important for the crew to remain circumspect in relation to the doings of the novice and give advice in case of problems or failure. For instance, when gaffing the fish, the deckhand was standing beside me with a long pole to secure the fish I kept losing with the hook—a practice that the experienced gaffer would do himself.

Image 4.3 The novice. The author working with the gaff hook to secure the fish while the lines are hauled in. It is no secret that the crew had great fun watching the novice miss a considerable amount of fish ('We have some experience')—in this case the deckhand would react rapidly and use a long pole equipped with spikes at the end (to the right of the author) to 'rescue' the valuable catch that had fallen off the hooks from the sea. To ensure the best quality, it is important to hit the fish with a single hard and precise stroke around the head so as not to damage the filets. When the lines hold only a few fish, the sailor in charge will often practise hitting the blank hooks with the tip of the gaff (Photo by AD)

In line with these observations, (Dreyfus & Dreyfus, 1984: 30) write that

> One must (…) abandon the traditional view that a beginner starts with specific cases and, as he becomes more proficient, abstracts and interiorises more and more sophisticated rules. It might turn out that skill acquisition moves in just the opposite direction: from abstract rules to particular cases. (ibid.)

It is important at this point to repeat that, independently from its level of mastery, skilful coping is *always* grounded in a temporal structure that bears reference to previous dealings. It is because of the temporality of man, as (Heidegger, 1962: §65: 403–408) remarks, that the novice is not completely at the mercy of a meaningless and alien environment, as he also has some sort of more or less advanced pre-understanding of what it means to be on a boat and fish for cod.

With Pálsson (1994) we can therefore say that the process of *enskillment* is rooted in the social- and natural environment, in which the novice enters a master–apprentice relation. Only on this basis can the novice learn to 'get his sea legs' in form of a bodily disposition. After all, it remains the privilege of the skilled expert to not only understand a given situation, but also to swiftly associate appropriate actions without engaging in any form of mental representations.[8]

The Care of the Skipper

In modern fishing, heavy machinery is deployed in a potentially rough and dangerous environment. Working out on the ocean not only requires a lot of skill and awareness of the surroundings, but also a strict set of routines, around which the fishing operation are structured. Hence, the skipper engages in care of the machinery and his surroundings to ensure that overall safety is maintained on board.

---

[8]Dreyfus and Dreyfus (1984) distinguish between four different stages of skill acquisition: the novice, the advanced beginner, proficiency and expertise.

In a similar way, the skipper carries the responsibility for the success of the fishing operation as such: the skipper decides not only when and under which conditions the crew will put to sea, but also strategy and fishing location. All these responsibilities require that the skipper has attained knowledge about crew, equipment and the natural environment. In a way, the role of the skipper can be compared to the daily routines and practices of scientists in particle physics, who lack direct access to their object of study and therefore engage in a set of routinised practices centred around control, observation and documentation of their detectors (Knorr Cetina, 1999). As Knorr Cetina (1999: 56) puts it, 'they substitute the care of the objects with the *care of the self* (italics in original). The care of the self, however, is not only an idiosyncratic quality of scientists, but a general structure that characterises our self-understanding (Heidegger, 1962: §41). Hence, all practices directed to equipment and others form the basis for the hermeneutics of self-understanding, which is first and foremost based on taken-for-granted ways and opinions concerning how 'one does things' (Heidegger, 1962: §27). In this sense, the care of the skipper compensates for the lack of access to and control over a highly volatile and unpredictable environment through a set of highly routinised practices, which form the basis for his individual knowledge and skill. In modern market-based fisheries, however, the care of the skipper is not limited to the fishing operation at sea, but also includes a set of routinised activities and practices on land.

If we follow Knorr Cetina (1999: 61), the care of the skipper can be analytically decomposed into three categories: (a) self-understanding, (b) self-observation and (c) self-description (see Image 4.4).

(a) *Self-understanding* concerns the understanding socio-technical components and processes of fishing. In order to ensure a successful and efficient fishing operation, it is important for the skipper to understand under which circumstances and season a fishing spot provides better results than another, which fishing spot produces better quality fish (also see Chapter 7), or under which conditions a certain type of bait will provide better results than another. Although the practices of the skipper are far from the meticulous research designs and controlled settings of a physics lab, skippers have hypotheses about the world that are based on experience and tested and fine-tuned in the daily practices of fishing.

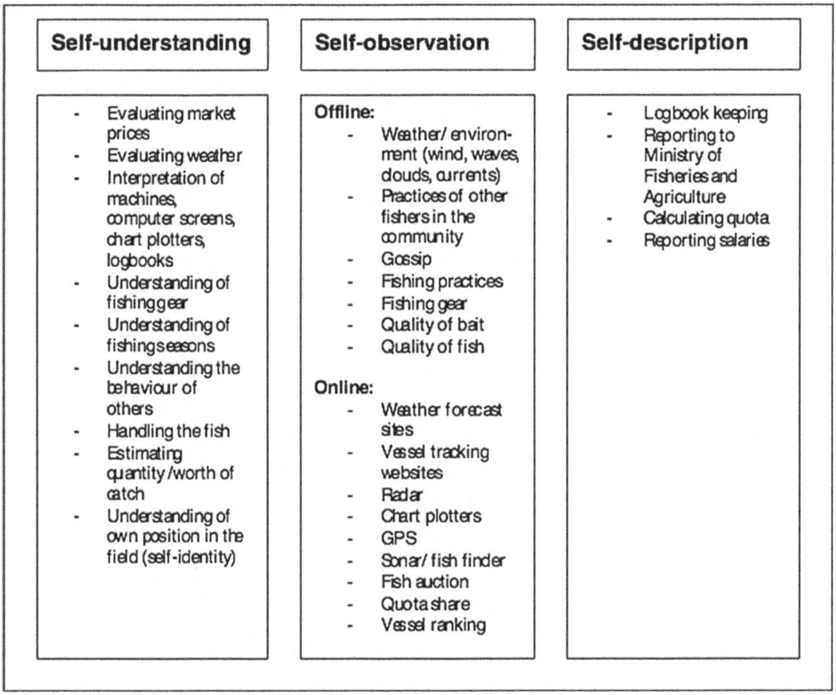

Self-understanding	**Self-observation**	**Self-description**
- Evaluating market prices - Evaluating weather - Interpretation of machines, computer screens, chart plotters, logbooks - Understanding of fishing gear - Understanding of fishing seasons - Understanding the behaviour of others - Handling the fish - Estimating quantity/worth of catch - Understanding of own position in the field (self-identity)	**Offline:** - Weather/environment (wind, waves, clouds, currents) - Practices of other fishers in the community - Gossip - Fishing practices - Fishing gear - Quality of bait - Quality of fish **Online:** - Weather forecast sites - Vessel tracking websites - Radar - Chart plotters - GPS - Sonar/fish finder - Fish auction - Quota share - Vessel ranking	- Logbook keeping - Reporting to Ministry of Fisheries and Agriculture - Calculating quota - Reporting salaries

Image 4.4 The structure of the Care of the skipper. (after Knorr Cetina 1999: 61)

For instance, when asking a skipper whose company was trying to save money by buying lower quality fresh bait if the different quality grading (A, B, C, and D) make any difference to the fishing he explains:

I have learned that if I have not the same bait as they have like in Bolungarvík, I cannot go near them, because the fish, it's just hurting my fishing, because the fish is not very happy [with] my bait if they have the other, you know, better [bait], so it's better for me to stay away from these guys. I know if I have the same bait, it's okay, I can be near the boats, but that is, that is a really really a big issue! If your bait is not 100 per cent, then you have to think: okay, it's much better for me to be on my own, no one around, then the fish have no other choice: bait – ah! [laughter]. (MXX)

Here, the skipper has developed a hypothesis about the relation between bait and catch over time, which in turn impacts his fishing practices. In a similar way, another skipper has a similar hypothesis about the age of his fishing line and the catch:

> Yeah, a new line is expensive, but it is expensive not to have it! Old line, we call it, it is in the sea it can be 'dead'. The sea has put so many things into the line and she is getting old, and you say she is 'dead' in the sea, so she is not moving, and a new line is moving and fishing more, fishing like 30 per cent more than the old one. (MII)

Although the percentage suggests some sort of empirical value, the fisher does not engage in exact testing or measuring the impact of new lines, but trusts his experience and folk knowledge. In a similar way, there seems to be no definite indicator of when to change a line: 'We see it, we just see it with our eyes' (ibid.).

Another important aspect of fishing is to understand when and under what conditions a fishing technique can be deployed. For instance, fishing with jigging computers is usually only effective over the summer months when the fish are active and the seas calm. In rougher seas, however, the drift is too strong to control the fishing lines, and over the winter months the fish are typically quite inactive and are better targeted with fresh oily bait on a stationary line.

Other important aspects of self-understanding include knowledge of the technical aspects of the fishing vessel, which is important in case of malfunction and emergency when out on the sea. Most importantly, understanding under which circumstances one can put to sea is one of the most important tasks of the skipper. This understanding, however, does not have the status of articulated knowledge, but is for the most part tacit knowledge that gains its meaning in its contextual embeddedness (see below).

(b) *Self-observation* concerns being aware of one's doings in one's own routines and practices. For this purpose, the skipper engages in a number of offline- and online practices in order to maintain the safety and success of the fishing operation. As shown above, these practices are first and foremost based on circumspection but may include more deliberate attention in situations of crisis or strategic decision-making. Offline practices include awareness of weather, cloud movements, currents, sea birds (as indicators for bait fish), the activities of other fishers (also through phone/wireless) and so on. Online practices include retrieving information of the weather forecast, vessel tracking websites, radar, sonar, GPS, market prices and so on (see also Chapters 6 and 8).

(c) *Self-description* concerns the documentation of a fishing trip. The skipper fills out a logbook after every trip, in which he notes data on day, fishing time and catch. Furthermore, the catch needs to be reported to the Directorate of Fisheries, which publishes all data on the catches of each vessel online. While this documentation is required by law, it also provides a platform for observing the rest of the fleet and planning future transactions and fishing operations. Especially with the construction of the quota market, fishers increasingly take care of calculations and paperwork not only in order to report their transactions to the Directorate of Fisheries, but also for keeping track of their own transactions. For this reason especially the 'online quota calculator' (see Image 4.5) represents an important device for making strategic decisions about fishing operations and market transactions and avoiding fees or losses, as the complex example of buying and renting makes clear:

> I make a paper, fax it to Fiskistofa [Directorate of Fisheries] and write in the computer how much I pay for the quota. Or if I rent it to [or] from me, sometimes we change if we, if we need more cod, or if we need more haddock. Sometimes we change, with the cod for haddock if we need them for the dry fish, or ocean catfish and I have to take care about if I rent more from me than to me … And I have to take care about [that], if I rent more from me than to me I don't [get] byggðakvóti [community quota] … so I will always have to take care about it in the computer to see in Fiskistofa [website of Directorate of Fisheries] how much is in the boat, can I change or… (MVIII)

To simplify this complex issue: the company needs to take care about how much they rent and lease, as they might violate the conditions for being part of the community quota programme, which is distributed by the Ministry of Fisheries to help fishing communities in decline and

Kvótategund	Þorskur	Ýsa	Ufsi	Karfi/gullkarfi
Úthlutun	180.666	353	3.816	250
Sérst. úthl.	0	0	0	0
Milli ára	-5.332	0	0	0
Milli skipa	0	0	0	0
Aflamarksbr.				
Aflamark	175.334	353	3.816	250
Afli	0	0	0	0
Aflabreyting				
Staða	175.334	353	3.816	250
Tilfærsla	0	0	0	0
Ný staða	175.334	353	3.816	250
Á næsta ár	27.100	53	572	37
Umframafli	0	0	0	0
Ónotað	148.234	300	3.244	213

Kvótategund	Langa	Blálanga	Steinbítur	Skötuselur
Úthlutun	299	9	397	1.094
Sérst. úthl.	0	0	0	0
Milli ára	0	2	0	164
Milli skipa	0	0	0	0
Aflamarksbr.				
Aflamark	299	11	397	1.258
Afli	0	0	0	0
Aflabreyting				
Staða	299	11	397	1.258
Tilfærsla	0	0	0	0
Ný staða	299	11	397	1.258
Á næsta ár	45	1	60	164
Umframafli	0	0	0	0
Ónotað	254	10	337	1.094

Image 4.5 The quota-calculator. This public online device gives skippers and quota owners an overview of their allotted and remaining annual quota shares and allows simple calculations with regard to future catches and transactions

requires quota-owners to fish off most of their own quota, rather than leasing it out. This, however, can be rather complicated when leasing in the system, which is based on so-called 'cod-equivalents' (*þorskigildi*), the anchor currency of the exchange system.

In this section we have pointed out that to compensate for lack of control and knowledge over the highly volatile environment in which fishing is taking place, fishers—in particular, skippers—engage in a set of routines and practices that allow them to gain knowledge and understanding of their environment. Based on this knowledge, fishers create what we might loosely call hypotheses and theories about the world, which are confirmed or falsified in the daily routines at sea. At the same time, the care of the self creates the basis for the skipper's self-understanding and individual skill. Today, many of these daily copings are not limited to the daily routines at sea, but transcend the spatio-temporal boundaries of the fishing vessel to encompass land-based practices. Depending on the complexity of the company, the routines and practices that are basic for the understanding of the skipper can also be differentiated. For instance, the unpopular paperwork is often taken care of by a third person, typically the mother, the wife or siblings who do not work at sea but are shareholders in the company, while bigger companies have professionals taking care of the paper work. It is, however, important for the fishers to be informed roughly about the 'business' and 'politics' that underlie fishing in order to plan their fishing operations according to the available resources. Many of the routines and practices that structure the fishing operation, however, are not based on explicit knowledge and documentation, but on unarticulated and incorporated forms of tacit knowledge, as the next section will show.

Fishing to the Limit

Ever since people started putting to sea, weather has been an important factor in maintaining the safety of the fishing crew. Especially small boats are quite vulnerable when it comes to rougher seas or sudden weather changes. Although vessels today are connected to a wide range of devices that synchronise skippers with a wide range of information on sea states and weather forecasts, skippers must be able to interpret and make decisions on their own.

Although the ITQ system has put an end to the so-called 'Olympic system',[9] in which fishers compete in a race for a finite amount of

---

[9]In the open coastal summer fisheries, however, the Olympic system still exists (see Chapter 3). Here, fishers compete for a limit amount of catch quota that are based on different zones, creating a higher incentive to fish in bad weather the more boats have registered for the annual season. The logic behind the system can be summed up as follows: if one fisher puts to sea, all need to follow in order to get their share of the allotted quota.

fish or total quota, fishers in some communities in the outer northern Westfjords even earn the reputation for having really 'tough' skippers. The reputation of the fishing community of Bolungarvík was thematised over and over again by many different people from different fishing communities during the period of fieldwork. The explanations of this behaviour, however, vary: while some interviewees just considered them 'crazy' and 'reckless', others considered them 'greedy' (MXIX, IX) and saw their way of fishing as irresponsible ('Something will happen' FN: 44). In contrast, skippers from Bolungarvík often showed pride in their reputation and tended to consider skippers from other communities as idle and not economically smart, as the following statement sums up:

> [They] are just always drinking coffee and relaxing – 'Oh it's good weather' – they just go home. And in Bolungarvík it is like, you know, when the weather is not good, then you get higher prices in the markets, it's all, the you fish a little bit less, but the higher price, so it's very good. (MI)

These different styles and attitudes towards rough seas makes clear that the definition of limits also depends on cultural framings and identities resulting from local rivalries that are diffused through different communities and may even vary from vessel to vessel. In this sense, fishers define and reproduce their limits depending on their socio-cultural embeddedness and economic means: a heavily indebted fisher will have more pressure to push his boundaries than one who has paid off his loans. In any case, it is obvious that skippers in any fishing community have a clear economic incentive to put to sea, as their income depends on the catch, and it is for this reason that fishers are prone to fish at their personal limits. But how are these limits defined in practice? When I asked a skipper whether he has a limit for putting to sea, he was not able to give a clear response:

> *AD*: What is your limit (for putting to sea)?
> *Skipper*: [tentative] The limit! It's very [flexible]! Can be here, it can be here... (IX)

The general ambiguity of defining a clear limit is confirmed by another skipper:

> *AD*: And just for you personally, is there a limit like wave height or something, where you say: Okay, I just don't do this (…)
> *Skipper*: Yes! Yes some, it's not not, you cannot count, write it down, say this is the limit… (MX)

The quotation points to the fact that the skipper knows very well for himself when to put to sea or not, but he is unable to verbally formulate a clear rule. Only when one pushes the skipper will he start to isolate different parameters:

> *AD*: Yeah, you check different parameters?
> *Skipper*: Yeah, check, yeah…
> *AD*: Like what do you check?
> *Skipper*: There's a wave buoy out the sea, you you can check that…
> *AD*: On the computer [on the website of the Icelandic Meteorological Office]?
> *Skipper*: Yes, if the metres [of the waves] are higher than two metres, three metres, then you maybe don't go. If the wind is more than 10 metres or something, it's ah, you have to, you have to check both, the wave high, and the wind, how strong is the wind…
> *AD*: And the interval between the waves probably…
> *Skipper*: Yes…
> *AD*: If it's long or…
> *Skipper*: And then you can, sometimes you can go behind the mountain, you have little bit…
> *AD*: Yeah, shelter… (ibid.).

Although the skipper has now suggested some parameters and some rules of thumb, the quotation still suggests that there is no clear rule that allows for exact calculation. Nor is there a device that, as an isolated artefact, tells the skipper what to do in a certain situation. Furthermore, the decision is also based on the environment of the fisheries. For instance, fishers have often highlighted that the fishers from the town of Bolungarvík also have better shelter for fishing, as their home port is situated at the mouth of a grand fjord and is therefore better protected than other ports which are either located directly on the open sea, or lie in small shallow fjords where fishing is considered unprofitable. However, a mountain or fjord that can provide shelter for one wind direction can turn out to be a dangerous choice in case of turning winds, and in any case, a fishing crew can also be 'locked in' to a fishing situation, as the following example makes clear.

> No, usually it is around, if the weather is not more [than] 10–15 metres [of wind speed per second], when we go out on the sea … Then we usually won't go out if it's, especially if it's over 15 metres, then we usually wouldn't go out in the, in the morning, but … If you go out in the morning and the weather is okay and then you start pulling the line in and suddenly you're hit, there comes a big wind, maybe up to 18 metres (per second), you try to, you always try to hang on, always try to finish it if you can! (MIX)

Nonetheless, if the skipper decides that the situation is getting too dangerous, the crew must cut off the line and head back to port.[10]

The examples have made clear that there is no clear parameter, rule or formula for decision-making. In order to decide whether to go to sea or not, Bjartur will wake up in the middle of the night, try to gain information on the activities of other boats, check the weather forecast and even check the movement of the clouds over the mountains:

> And if you look into the sky and you see the movement of the skies [clouds], are they moving fast or are they moving slow, they are moving slow now, so that should not be so strong wind outside. But you can also [see] on the left hand, here over the mountain, you can see the sky is very dark and that tells you – this is not very beautiful to see when you're going

---

[10]Although fishers try to avoid losing lines, there is still a chance to retrieve a line when the weather conditions have calmed down again. To do so, the skipper can navigate back to the exact spot based on GPS data and drag a modified anchor over the bottom of the spot in hope of entangling the fishing line in the device.

to see this, but on the other side you can see it's a lot like this, it's a different, it's a cloud there, if you look there, it's better, you can see a difference,
just by the left side of the fjord, on the right side it's very bad ... but you
can also look at the mountain and I think the wind is coming from the
top and going down, you cannot see anything now, but sometimes you
can see it and you can listen if you put your head out the window if the
sea is broken. (MII)

Bjartur makes clear that although forecasts nowadays are rather
accurate, experienced skippers engage in the complex art of 'reading
nature' just as their ancestors did in pre-modern times. To do so, skippers not only rely on visual observations, but also on acoustic signals
that stem from their closer environment. Hence the decision whether
to go to sea or not is based on circumspective awareness and the bodily
relation to a specific situation that makes up the equation of whether to
put to sea or not, rather than being based on some explicit rule or limit.
Hence, the limit itself must be seen as a social construction based on the
contingent situatedness of local knowledge, socio-material practices and
entanglements of the fisher in the world of fishing. These limits based
on experience and skill at sea and of course can be proven false when
skippers try to push their vessel beyond its capacity. Although safety
measures have improved significantly throughout the years and digital
technologies allow for all sorts of information on sea state, wave speed
and weather forecast, unpredictable rapid weather changes may surprise
a skipper when out in the Arctic Ocean. In case a vessel gets into trouble, however, it can usually still rely on a wide range of assistance from
other larger vessels that provide shelter and the coastguard. Although
serious sea accidents have become a rather rare event, new economisation and new technologies have reconfigured risk-awareness in the small
boat economy, as we will see in Chapter 6.

We have now seen that evaluations of the environment for safety concerns are not based on some form of quantifiable objective indicators
that allow for clear calculations, but on tacit knowledge and the particular circumstances of a specific situation. These circumspective evaluations of the environment are a form of skilful coping, which is based on
incorporated forms of explicit knowledge that is gained and reproduced
in the daily routines and entanglements at sea. For this reason, not only

the assassement of the immediate environment, but also reputation in the community, economic incentives and pressures have a big impact on a skipper's conception of risk. Thus, the social context co-constitutes the interpretative boundaries in which skippers base their judgements and decision-making. The next section will elaborate further on the socio-technical nature of fishing.

Fishing with Others

It is certainly correct that the idea of the a-social outlaw fisher proves wrong because he is embedded in rural production networks (Acheson, 1988; Barnes, 1954). This section argues even more strongly that the practice of fishing itself is inherently social. In order to understand the social nature of fishing, let us return to the beginning of our fishing trip.

When leaving port, it strikes me that the skipper is talking constantly to other fishers on the wireless at sea. Intrigued by this, I ask him whether he sees the other fishers more as competitors or colleagues. After a short pause and obviously puzzled by the question he responds: 'It's strange, it's both' (FN: 76).

Information about current fishing activities is key to success and skippers engage in all sorts of domestic networks and social institutions such as the local swimming pools or coffee places. At sea, however, modern communication technology (also see Chapters 5 and 8) is key to success not only in finding a free stretch for putting out the fishing lines, but also for gaining information on current fishing activities, as the following example from a jigging trip in the summer makes clear.

After the first drift remains rather slow and below the skipper's expectations, the skipper decides to call on a trusted skipper located about 10 miles away from us. To gain information about fish activity in that area, we were invited over to fish next to him. After a first busy drift at the new spot, the distance between the vessels was so close that we could even communicate orally and threw jokes at each other while working the fish coming up on deck (FN: 95–96). The little episode shows that the social organisation of modern fishing exists not only in networks and institutions that entangle each individual fisher to the domestic world of production, markets and politics. Rather, the practice of fishing itself is mediated through the local ties and socio-technical networks, through which information flows.

The field examples altogether show that, despite local rivalry, fishers depend on each other and rely on trusted relations and network ties to gain access to information on current fishing activities. But it is not only for this reason that it is wrong to conceptualise fishers as individualistic loners that tend to hide information from others.

Even the most 'solitary' fisher putting to sea is always already related to the social origin of the devices that make up the floating workshop (Heidegger, 1962: §15): the hull of the fishing vessel refers to the people who built it; the fishing lines refer to the baiters in the community; the computer software on the chart plotter refers to its programmers and so forth.[11] Merely unravelling all sorts of relations between fishers and their socio-material environment, however, runs the risk of being a rather descriptive endeavour. If we follow Heidegger (1962: §15) further, sociality is not first discovered in the process of interaction (Strauss, 1978) or based on integration in a normative system (Parsons, 1951). Rather, human beings are inherently social and by definition entrenched in a meaningful shared world with others. Thus, fishing, like any other form of skilful coping must itself be understood as a socio-technical practice, that is, a set of routinised activities that are grounded and reproduced in a shared and always already discovered meaningful and historically contingent world.

The existential analysis has pointed out the social nature of fishing as a practice. Although fishing is by definition a social practice, as shown above, activities must be coordinated between fisher and technology and other crewmembers, respectively. But how do small boat fishers coordinate their practices at sea?

---

[11]Heidegger (1962: 70) also remarks that the practical use of equipment also refers to its material and origin in 'nature'.

Beyond Formal Hierarchy

Large naval vessels are traditional examples of *total institutions* par excellence; that is, 'a place of residence and work where a large number of like-situated individuals, cut off from the wider society for an appreciable period of time, together lead an enclosed, formally administered round of life' (Goffman, 1961a: IX). While this definition may also fit some larger industrial fishing vessels, in which clear cut-hierarchies and chains of command structure coordination on board (Barnes, 1954), empirical observations suggest that life on small boats takes place on an almost exclusively informal basis. Hence, social organisation on small boats neither follows an 'over-all rational plan' (Goffman, 1961b: 6), nor is it based on a clear-cut division between a 'large managed group' or 'staff' (ibid.: 7). Space on board ship is simply too narrow to even allow for spatial separation of the kind found on cargo ships (Sandberg, 2014)[12] or larger fishing vessels. Instead, all spaces—the deck, the wheelhouse and the small space with two bunks and a little cooker and a microwave for the preparation of simple meals[13]—are shared by the crew regardless of rank or social position. Although the skipper is usually respected as the main authority having the last word on every decision and responsibility for navigation, there are no uniforms or any other visible symbolic distinctions made between the crew and the skipper, who both wear more or less the same working clothes on deck. Verbal communication is highly informal on board and fishing strategies and schedules are usually openly discussed among the crew. But how are practices coordinated when hierarchies are rather flat in an industrialised environment that requires smooth and fast decision-making?

A common longliner classified in the Icelandic small-boat system typically consists of 2–4 sailors depending on vessel size and quota share. Within the crew, different roles ascribe a set of expectations to each

---

[12]For instance, Sandberg (2014: 102–103) reports the spatial separation of different ethnic groups on cargo vessels during coffee breaks.

[13]Even smaller or older vessels usually lack bunks and usually have no cooking space onboard.

crewmember during the fishing operation.[14] Typically, the skipper is in general charge of the fishing operation and decides on schedule, strategy and fishing grounds and takes care of all navigation-related issues at the wheelhouse, while a deckhand is in charge of all deck-related activities, such as loading the boat, putting out the lines and working the fish.

Modern small-boat fishing has some structural similarities to *heterar-chies* (Stark, 2009),[15] forms of organisation that are not based on a top-down hierarchy, but on multiple, equally ranked units of command. The complete decoupling of work units, however, would be counterproductive if not dangerous in the context of heavy machinery and rough seas, and no signs of different orders of worth among different ethnic groups can be found as reported by Sandberg (2014) in the case of large cargo vessels. Hence, a successful and efficient fishing operation is rather based on the frictionless coordination of tasks that requires attentive observation of the environment and mutual understanding of each other's doings, for instance when setting the lines (see Image 4.6). Hence, coordination on small boats requires circumspection that enables routinised and smooth coordination of activities with others in a highly technisised and potentially rough work environment.

Circumspective coordination designates a mode of highly routinised and flexible skilful coping that allows the coordination of multiple activities in a highly uncertain environment. Routine in this context refers to taken-for-granted roles and ways of 'doing things'; that is, the daily routine of repetitive practices with equipment and other people that structure the overall framework of a given social order. Flexibility refers to the versatility of roles and practices with regard to the exigencies of a given situation. Thus, coordination of practices on a small fishing boat is not so much based on static role structures and formalised chains of command, but on a dynamic set of flexible roles that provide

---

[14]Depending on the vessel and size of the crew, activities of course will be more or less differentiated, and some of the bigger small boats consist of a skipper, a marine engineer and one or two deckhands.

[15]From the wider perspective of a producer market, fishing vessels has always been embedded in a heterarchical structure as producers have little to no direct influence over *where* and *how* a fishing vessel fishes—even if contracted or vertically integrated into a company.

Image 4.6 Setting the lines. After almost four hours' sailing, at around 3:45 am the skipper wakes up the deckhand to prepare everything for releasing the lines. First, the buoy is prepared for release; a heavy lead as anchor is attached to a line, to which the first fishing line of the string is attached. The system has retained its core simplicity, 'not rocket science', as the deckhand puts it, but mechanisation has added a new quantitative dimension to this fishery. Each of the 32 bala is put under a special appliance at the stern of the vessel. Now, the skipper has to manoeuvre the vessel much more slowly so that the line can be released in an orderly fashion over the fishing spot. When a bala is emptied, the next one will start running and the deckhand will exchange the empty bala with a full one until the desired stretch of water is covered. It is very important that the baiter has tidily rolled up the fishing line into the bala to avoid entanglement and ensure an accurate release. At the end of each line, the new line is attached, so that all lines create one single longline. The line, however, is not laid out in one single stretch, but in parallels (in this case running north to south). After the first stretch is set, the deckhand will merely signal orally that the skipper can now turn the boat for the second stretch. Releasing all lines will take us about one and a half hours. After this work is done, the deckhand will come back into the wheelhouse and prepare himself for a long day on deck: 'Now you have to eat a lot, it will be 6–7 hours work then' (Photo by AD)

orientation for the changing situational dynamics in which small-boat fishing takes place, as the following examples make clear.

After a hearty meal in the wheelhouse around 6 a.m., we start hauling in the lines with the help of a special grid winch machine in the front part on the backboard side of the vessel. First, the buoy is picked up, and after the first 'blank' line is hauled in, the first fishing line will soon break the surface. Although the fishing vessel is equipped with heavy machinery, a large part of the labour out on deck is based on artisanal and physically demanding labour, in which the skipper does the same work as the deckhand. The organisation of labour is fairly simple and flexible: one gaffes the fish by the winch to make sure the fish slip safely on board into the metal storage box, while the other waits to cut the throat of the fish and throw them through a metal tube underneath the deck, where boxes with ice and sea water make sure the fish can bleed out and is cooled down immediately—a practice that will turn out to be of crucial importance in the context of economising (see Chapter 7).

On a small fishing vessel, however, the amount of tasks outnumbers the crew, which needs to rely on teamwork and improvisation: depending on the amount of lines, the crew will rotate between the tasks of gaffing the fish, cutting the fish's throats for bleeding before forwarding them to the hold, shovelling ice in the hold (see Image 4.7), ordering the boxes and having an eye on the wheelhouse and the movement, speed and direction of the vessel for the next 6–7 hours.

Hence, informal chains of command, flexible roles and rotation not only bring relief to the long hours of hard, monotonous and repetitive work, but also allow for more flexibility on deck. Short brakes for consuming a dose of snuff or the satisfaction of other needs are usually executed without stopping the grid winch and usually when only a few fish are coming up with the lines. The winch and the fishing line, however, must be under the constant surveillance and control of at least one sailor as long as lines are hauled up. Not only would a missed fish mean an economic loss; a tangled or snagged line may even cause serious injury or even disaster. In cases of danger and mechanical failure, however, the crew needs to improvise and sense rapidly what has to be done in a given situation without waiting for a clear command from a superior or explicit coordination of tasks (for instance when the line gets tangle, as shown above). For this

Image 4.7 Teamwork. Out on deck, the crew works together as a team. Skipper and deckhand usually switch between different tasks, mainly gaffing the fish, cutting the fish, checking on the load of the boxes in the hold and making sure that the fish are cooled down with ice. In this case, the deckhand has switched from cutting the fish during a period of blank lines to shovelling ice into the hold without a direct order or command by the skipper, who is watching the line. To make the long, monotonous and physically demanding work and the lack of rest on a heavily rolling boat more bearable, the crew sings along with the songs coming from the wheelhouse's radio or enjoy a good dose of strong snuff (Photo by AD)

reason, the circumspective coordination of tasks implies that crewmembers are aware of their work places and sense when to shovel more ice into the fish storage, cut the fish, take care of the winch or when it is appropriate to take a break. Thus, circumspective coordination is not merely some formal organisation of consciously executed tasks by different roles, but a skilful coping strategy based on incorporated knowledge that responds to the structural needs of a narrow workplace situated in a potentially dangerous and ever-changing environment of heavy machinery and unpredictable seas.

Conclusion: Fishing as Socio-Technical Practice

This chapter has attempted to take the reader on board a long lining vessel to gain an 'inside' perspective on the world of modern small-boat fishing. By joining a fishing crew at sea, it has become obvious that the meanings and practices of today's coastal fisheries have little to do with the much romanticised world of open wooden small boats fishing from yesteryear. Rather, small boat fisheries, in particular modern longlining, has grown to an industrial off-shore activity involving heavy machinery, multiple kilometres of fishing lines equipped with thousands of hooks capable landing multiple tonnes of fish each trip. Part of this transformation can be traced back to the economisation of the fisheries itself, which has not only translated fish stocks into a manageable resource and commodity, but has also transformed the world of fishing into an increasingly efficient year-round activity.

Instead of being reduced to mindless cogs in a relentless 'harvest machinery' (Johnsen, 2004), however, the ethnographic material reveals that modern small boat fishing is still based on skill and tacit knowledge that enable fishers to cope and swiftly adjust their practices to a rough, potentially dangerous and ever-changing environment. For doing so, fishers are always already in a disclosed meaningful world that forms the basis for their daily coping that is structured around circumspective, that is mostly pre-reflexive bodily involvement with equipment and others. Moreover, our observations suggests that this 'skilful coping' with the environment is first and mostly a routinised, team work-based form of flexible coordination that lies before formal chains of command and decision-making. In contrast to pragmatist accounts that focus mainly on the creativity of daily problem-solving (Joas, 1992), however, the Heideggerian analysis of being-in-the-world of fishing points at a more general relation of humans with their world that not only provides a richer empirical description of pragmatic coping (Blattner, 2008), but locates the relation of daily coping within a historically contingent web of socio-technical relations and practices.

In order to fully fathom how and to what extent modern small boat fishing has been transformed by its new socio-technical entanglements, the following chapter will delve deeper into the relation between modern technology and human coping.

References

Acheson, J. M. (1988). *The Lobster Gangs of Maine*. Hanover: University Press of New England.

Barnes, J. A. (1954). Class and Committees in a Norwegian Island Parish. *Human Relations, 7,* 39–58.

Blattner, W. (2008). What Heidegger and Dewey Could Learn from Each Other. *Philosophical Topics, 36*(1), 57–77.

Bourdieu, P. (1977). *Outline of a Theory of Practice*. Cambridge: Cambridge University Press.

Dobeson, A. (2018). Between Openness and Closure: Helmuth Plessner and The Boundaries of Social Life. *Journal of Classical Sociology, 18*(1), 36–54.

Dreyfus, H. L. (1991). *Being-in-the-World: A Commentary on Heidegger's Being and Time, Divison I*. New Baskerville: MIT Press.

Dreyfus, H. L. (1993). Heidegger's Critique of the Husserl/Searle Account of Intentionality. *Social Research, 60*(1), 17–38.

Dreyfus, H. L. (2004). Merlau-Ponty and Recent Cognitive Science. In C. Carman & M. Hansen (Eds.), *The Cambridge Companion to Merleau-Ponty* (pp. 129–150). Cambridge: Cambridge University Pres.

Dreyfus, H. L., & Dreyfus, S. E. (1984). From Socrates to Expert Systems: The Limits of Calculative Rationality. *Technology in Society, 6*(3), 217–233.

Foucault, M. (1970/2002). *The Order of Things: An Archeology of the Human Sciences*. London: Routledge.

Goffman, E. (1961a). Introduction. In E. Goffman (Ed.), *Asylums. Essays on the Social Situation of Mental Patients and Other Inmates* (pp. IX–XIV). New York: Random House.

Goffman, E. (1961b). On the Characteristics of Total Institutions. In *Asylums: Essays on the Social Situation of Mental Patients and Other Inmates* (pp. 3–124). New York: Random House.

Heidegger, M. (1962). *Being and Time*. New York City: Harper Collins.

Heidegger, M. (1983). Gesamtausgabe. II. Abteilung: Vorlseungen 1923–1944. Band 29/30. Die Grundbegriffe der Metaphysik. Welt - Endlichkeit - Einsamkeit. Frankfurt am Main: Vittorio Klostermann.

Hollan, J., Hutchins, E., & Kirsh, D. (2000). Distributed Cognition: Toward a New Foundation for Human-Computer Interaction Research. *ACM Transactions on Computer-Human Interaction, 7*(2), 174–196.

Hutchins, E. (1995a). *Cognition in the Wild*. Cambridge, MA: MIT Press.

Hutchins, E. (1995b). How a Cockpit Remembers Its Speeds. *Cognitive Science, 19,* 265–288.

Joas, H. (1992). *Die Kreativität des Handelns.* Frankfurt am Main: Suhrkamp.

Johnsen, J. P. (2004). The Evolution of the "Harvest Machinery": Why Capture Capacity Has Continued to Expand in Norwegian Fisheries. *Marine Policy, 29,* 481–493.

Knorr Cetina, K. (1989). Spielarten des Konstruktivismus. *Einige Notizen und Anmerkungen. Soziale Welt, 40*(1/2), 86–96.

Knorr Cetina, K. (1999). *Epistemic Cultures: How the Sciences Make Knowledge.* Cambridge, MA: Harvard University Press.

Knorr Cetina, K. (2001). Objectual Practice. In T. R. Schatzki, K. Knorr Cetina, & E. v. Savigny (Eds.), *The Practice Turn in Contemporary Theory* (pp. 176–188). London and New York: Routledge.

Latour, B. (2005). *Reassembling the Social: An Introduction to Actor-Network -Theory.* Oxford: Oxford University Press.

Merleau-Ponty, M. (1945/2012). *Phenomenology of Perception.* Abingdon, Oxon: Routledge.

Pálsson, G. (1994). Enskilment at Sea. *Man, 29*(4), 901–927.

Parsons, T. (1951). *The Social System.* Glencoe: The Free Press.

Reckwitz, A. (2002). Toward a Theory of Social Practices: A Development in Culturalist Theorizing. *European Journal of Social Theory, 5*(2), 243–263.

Sandberg, C. (2014). *On Board the Global Workplace.* (PhD dissertation). Stockholm University, Stockholm.

Schatzki, T. R. (1996). *Social Practices: A Wittgensteinian Approach to Human Activity and the Social.* Cambridge: Cambridge University Press.

Schatzki, T. R., Knorr Cetina, K., & Savigny, E. v. (2001). *The Practice Turn in Contemporary Theory.* London and New York: Routledge.

Schütz, A., & Luckmann, T. (1975/2003). *Strukturen der Lebenswelt.* Konstanz: UVK Verlagsgesellschaft/ UTB.

Stark, D. (2009). *The Sense of Dissonance: Accounts of Worth in Economic Life.* Princeton: Princeton University Press.

Strauss, A. L. (1978). *Negotiations: Varieties, Contexts, Processes, and Social Order.* San Francisco: Jossey-Bass.

5

Enframing the Sea

*Improvements in the technique of rowing boats do
not lighten the labour, which is required of those
who desire to hold a good position in a great race.*

Alfred Marshall, Industry and Trade (1919)

While hanging out at the wheelhouse shortly after we started our long
journey back to port after a long day of work at sea, I cannot hide
my excitement about all the computer screens in the wheelhouse (see
Image 5.1):

AD: This place reminds me of the cockpit of a space shuttle!
Deckhand: Oh, this is nothing!
Skipper: You should see the new bigger small boats of the fleet, they have
even more screens with 3D technology - they can really see more!
AD: But do you really catch more fish?
Deckhand: Oh yes!
Skipper: If you know how to use it! (FN: 79–80)

Today, the wheelhouses of small fishing boats are packed with observational *scopes* that a few years ago would have been found only in

© The Author(s) 2019, corrected publication 2020
A. Dobeson, *Revaluing Coastal Fisheries*,
https://doi.org/10.1007/978-3-030-05087-0_5

Image 5.1 The wheelhouse as multi-scope. Modern navigation technology broadens the perspective of the skipper to not only observe other vessel activity in the closer environment (radar to the lower right), but also to gain information on the sea bed structure (sonar in the upper middle) and water temperature in different areas (monitor, lower middle). Furthermore, modern navigation software allows skippers to store positioning and vessel tracks based on GPS data (monitor, lower left) (Photo by AD)

coastguard ships and large fishing trawlers. Our vessel is equipped with no fewer than six (!) additional screens and chart plotters revealing a broad range of information from below the sea and the vessels surrounding environment (i) a radar displaying nearby vessels; (ii) a sonar/fish finder displaying current depth, sea bed structure and fish underneath the boat; (iii) a GPS; (iv) a special marine navigation software displaying the vessel's past and current positions and other real-time vessel movements at sea based on GPS data and sonar; (v) a second screen used for displaying either an extension of (iv), or displaying the Icelandic Meteorological Office's weather forecast, in particular the wind forecast, the website of the Directorate of Fisheries or other websites not necessarily used for navigational purposes; and (vi) a screen in the middle of the wheelhouse mostly used for recreational use while sailing, such as streaming movies via the vessel's satellite internet connection and a small screen to the upper right connected to a camera for monitoring the engine room. Especially the latter has proven to be a life-insurance in today's economised and intensified world of fishing, as the skipper explains: 'This is very important, especially if you are going long way out'. And in fact, I witnessed several times during my time in the field that vessels had to be towed back to port by other vessels due to mechanical breakdown of the engine. As the skipper explains to me with regard to the camera in the engine room, the worst nightmare of any sailor, however, is fire on board a vessel—an incident that had just occurred a couple of days before on a boat from a neighbouring community. Luckily, the fire could be extinguished and the boat towed back to port without anyone getting injured. The skipper remarks, however, that using computer screens require skill and do not per se lead to safer and more successful fishing operation. Chart plotters, sonar and GPS have to be understood not only in their technical aspects, but must be related to the socio-technical practices of fishers coping with a rapidly changing environment in which modern fishing takes place.

Up to this point we have described the practice of fishing in terms of the skilful copings and circumspective routinised practices around in which the world of fishing is constituted and reproduced. This perspective is based on a rather instrumental conception of technology as mere tool that is deployed 'in-order-to' achieve a certain end (see Chapter 4). Our little episode above, however, raises a more fundamental question about modern technology itself: how can we understand this constant desire to 'see more'? And what does it imply for the economised world small-boat fishing?

In this chapter we will see that modern navigation technology can be understood neither as something that determines social action, nor by simply reducing it to a means that serves the fulfilment of human ends. Rather, modern technology is a form of *enframing* (Heidegger, 1977) of the world that creates new pathways for social practices by bringing forth new forms of meaningful relations. From this perspective, digital navigation technology enframes the sea by linking and synchronising fishers with the movement of fish stocks, consequently changing the temporal and spatial orientation of the world small boat of fishing itself.

Modern Technology as Enframing

At least since Marx has the role of technology gained a prominent role as major driver of social change in the social sciences.[1] Debates about how exactly technology 'acts' upon society, however, have been widely contested ever since. Whereas technological determinists tend to understand technology as something 'hard' and 'objective' independent of society itself, others have pointed out that technology is fabricated in a socio-historical context, rather than simply imposing itself on society (MacKenzie & Wajcman, 1999a). From this perspective, technology is a *human doing* rather than a deterministic *doing human*. Thus, the evolution of scientific innovations and technological progress have to be understood in their historically contingent context, in which not necessarily the 'best' or 'most efficient', but the socially most viable solutions and devices are established with regard to historically evolved social constellations and power relations.

While both determinist and constructivist positions represent forms of reductionism that give primacy over either technology or society, later accounts have revised this argument by pointing at the structural symmetry and co-constitutiveness of technology and society.[2] This perspective has opened up some fruitful insights into the role of human/non-human relations for the construction and stabilisation of controversies around socio-technical arrangements (e.g. Callon, 1998; MacKenzie, 2009). In line with the works of Donna J. Haraway (1985), which challenge human–nature–technology distinctions, a number of scholars have pointed out processes of 'cyborgisation'[3] of the fisheries that blur the lines between nature and culture (Holm, 2001; Holm & Nolde Nielsen, 2007; Johnsen, Holm, Sinclair, & Bavington, 2009).

---

[1] For an overview of the role of technology in Marx' writings, see MacKenzie (1984)

[2] In the second edition of *The Social Shaping of Technology* (MacKenzie & Wajcman, 1999b), the editors correct this yet again one-side by taking into account the actor qualities of technology and artefacts for the co-constitutiveness of technology and society (ibid.: 23–24).

[3] In the context of Haraway's (1985) seminal text, the 'cyborg' refers to a political anti-hero that is ought to challenge hegemonic essentialist conceptions of gender, humans, technology and society.

While these accounts do provide good 'macro' descriptions of the emergence and stabilisation of actor-networks over time, little attention has been given to the relation between technology and practice. So how can we understand the mutual relations between human coping and technology?

Decades before the field of Science and Technology Studies was established as a field and the role of technology and material artefacts gained wide acceptance among social scientists, Martin Heidegger (1977) criticised mainstream conceptions of technology for reducing technology to means–end relations, which is inherent in the classical anthropological definition that sees modern technology as a toolbox serving the fulfilment of human needs.[4] For the later Heidegger (1977), modern technology was no mere means or just a certain way of doing as his early writings indicate, but a new way of how humans relate to themselves and to their world. Rather, for Heidegger, technology as such is a way of *bringing-forth*, that is a *revealing* of meaning in the world, rather than a mere tool used to alter the world such as a hammer or a fishing line. With the emergence of the exact sciences, however, this way of revealing has changed radically and become what Heidegger refers to as *en-framing* (*Ge-stell*) that radically alters the ways in which we relate to our world as such.

Understanding technology as enframing makes clear that it is not only the 'cognitive frames' as such that are of importance for the understanding of technology. Rather, the wording makes clear that it is the active process of framing itself that come into focus. Hence, en-framing implies that something is put in place by means of a frame, like a piece of art that is mounted to a wall. The wording also makes clear that it is not primarily a human being who puts the artwork into its place, but the frame as such. In this sense, the frame creates a forum in which the piece of art reveals meaning relative to a certain epoch in time.

---

[4]Heidegger traces the etymology of the word 'technology' back to its primordial meaning, which derives from the Greek word *technikon*. Accordingly, *technikon* implies everything that belongs to a technique (*technē*) (ibid.: 12): '*technē* is the name not only for the activities and skills of the craftsman, but also for the arts of the mind and the fine arts' (ibid.: 13).

With the dawn of the exact sciences and machine power, however, technology no longer brings forth meaningful objects and goods, but becomes a way of revealing oriented towards the challenging of 'nature' as potential resource that can be measured, extracted and stored:

> The revealing that rules throughout modern technology has the character of a setting-upon, in the sense of a challenging forth. That challenging happens in that the energy concealed in nature is unlocked, what is unlocked is transformed, what is transformed is stored up, what is stored up is distributed, and what is distributed is switched about ever anew. (Heidegger, 1977: 16)

To clarify what Heidegger means by understanding modern technology as a challenging-forth of nature, it is useful to recall an example from the text that is based on the difference between an old wooden bridge and a hydroelectric plant on the Rhine. Whereas the old wooden bridge was built into the river to enable the crossing of the river, the relation of the river to the hydroelectric plant is reversed:

> The hydroelectric plant is not built into the Rhine as was the old wooden bridge that joined bank with bank for hundreds of years. Rather the river is dammed up into the power plant. (ibid.: 16)

Thus, the plant materialises the setting-upon of modern technology, which regulates the water flow while transforming it into a disposable *standing-reserve* (*Bestand*) ready for utilisation (ibid.: 16). It is in this sense that the way of revealing of modern technology is a *setting-upon*, which reveals a distinctively modern condition that is grounded in the exact sciences.[5]

By conceptualising modern echnology as and active enframing we can now understand better understand the ways in which our small boat skipper relates to his world: while in pre-modern times, open rowing

---

[5]For Heidegger (1977: 16), the essence of modern technology was prepared long before the dawn of its age by the rise of physics 'for already in physics the challenging gathering-together into ordering revealing holds sway'.

boats were at the mercy of the current, the tides, the waves and the movement of the fish, the world of modern fishing tends to reverse the process: with the advent of motors, the sea has become more and more a subject of order that can be mastered by machine power, turning the fish more and more into a standing reserve that can be mastered and controlled. In this sense, the relationship between the fishing fleet and the ocean is somehow reversed: rather than the fishing fleet being challenged by the mercy of the sea, the sea is challenged by the mercy of the fishing fleet, which runs year-round operations with ever more efficient vessels. It is also in this light that we have to understand modern resource management, which is not fundamentally changing the relation between the fishing industry and the ocean as a resource; it is invented to conserve the ocean as a standing-reserve that can be managed and controlled by scientific means. This, however, does not mean that modern technology is something separated from human practice, but it does not 'happen exclusively in man, or decisively through man (ibid.: 24), as we find ourselves always already in a world that is largely constituted by modern technology. In this sense, not only the ocean and the fish, but also the fisher is enframed by modern technology. Modern technology thus is an enframing of practice that reveals the world as a-historical, objective and quantified space that can be managed and controlled.

Hence, in contrast to the olden days, when man had to cultivate a field with the seasons or live with the tides of the sea, modern technology makes crops and fruit available for supply and distribution all year round by means of modern communication and transportation technology. The challenging-forth of nature is the constant accessibility of 'nature' as a resource, which is pushed to the extreme by extracting all sorts of energy from nature and storing it when the market as central form of coordination has a demand for it. Large modern fishing vessels, which can put to sea in strong winds and metre-high waves allow the stabilisation of international markets by providing a constant supply of raw materials. Freezing technology today allows processors and sellers to stack and store fish for international markets. It is this constant availability that characterises modern technology and the temporal relation of humans with 'nature'. This challenging forth also holds true for daily communication and transportation technologies which redefine the spatio-temporal boundaries across different social worlds—an aspect that will later turn out to be crucial for our understanding of the 'small boat revival'. First, however, the next section will therefore point out how the emergence of a new form of digital enframing has reconfigured the temporal and spatial orientation of the skipper by revealing the underwater world beneath him, hence enabling to synchronise his practices with moving fish stocks and the ever-changing environment of the sea.

Fishing with Scopes

What Heidegger could not foresee was how digitalisation would add a new qualitative dimension to modern technology. Today, digital technology has redefined the ways we live by linking and synchronising the world in one all-encompassing digital network, which allows us to exchange information in a split second and redefines business relations as a pioneering field of a globally linked 'network society' (Castells, 2001). Especially digital 'scopic media', that is instruments 'for seeing and observing' (Knorr Cetina, 2003: 8) such as chart plotters and computer screens mirroring digital information play a crucial role in redefining the successive 'pipe-logic' of networks towards a more flow-like structure of information exchange (Knorr Cetina, 2003). In other words, modern information and communication technologies enframe and challenge social relations towards a globally synchronised whole by means of computer screens and mobile phone displays. The same development can also be observed in the fisheries, in which the temporal and spatial boundaries between the fish and the fisher have diminished.

During my period of fieldwork in the summer of 2014, I witnessed a new development in the coastal fisheries, in which skippers took their crew and their state-of-the-art long line vessels far out, 40–60 nm off shore, breaking one record after another with single-haul landings well over 20 tonnes.[6] According to the fishers at the docks, this is a new

---

[6]At this time, the maximum tonnage in the hook and line system was only 15 GT. I witnessed some of these 'record hauls', in which the vessel was lying very deep in the water while approaching the harbour very carefully and slowly.

strategy, which was enabled by the faster and stronger engines deployed on the newer, slightly larger coastal vessels used for longlining. While it is certainly new that small boats are no longer limited to fishing in the 12 nm coastal zone,[7] modern information and navigation technology is also playing an important role in disclosing new areas at sea. This becomes clear in the following exchange with a young fisher working as engineer and deckhand on one of these boats:

AD: I heard about people catching so much fish on few boats. Does that have something to do with the way they are fishing, or is there only a lot of fish right there?

Fisher: Yeah, we do it differently now than the last years. The sea is hot here and cold here, you know [pointing towards a monitor with a heat map, where one can see different zones coloured from dark blue to red] (…)

AD: Yeah, but there are different kinds of areas in the sea with different water temperatures.

Fisher: Yeah, and the fish is in the water with the difference (…) I can just show you here…

AD: Ja…

Fisher: Here's Bolungarvík, we've been here on the long liners, here [pointing at fishing spot]. See, there's the cold sea coming here, putting the fish here in [between warm and cold water]…

AD: Okay…

Fisher: And there's the hot sea, and the cold sea is pulling in here, so it's get a lot of fish in this, in this small area, and we are crossing the lines here [between the hot and the cold sea]. (MXVI)

The fisher explains how recently fishers from the area have discovered that fish tend to congregate in a small area about 40 nm from their homeport in the summer. Although the fisher has only a vague idea why the fish congregate between the colder and warmer zones, he nevertheless knows that this *is* the case and that it is important for ensuring the

---

[7]In Iceland, the fishing grounds within the 12 nm zone are exclusively reserved for boats classified in the small-boat system.

best possible placement of the lines. Different from the seabed structure, which changes rather slowly, the strong currents of the sea can change the temperatures in different zones rapidly. For this reason, it is important for skippers to gain knowledge of their environment through socio-technical networks and keeping up with the newest technological development to ensure the best possible spot for their fishing lines to fish off their quota as efficient as possible.

In general, these digital scopes are meshed into the care and circumspection of the skipper (see Chapter 4), who not only monitors the course and state of the vessel, but also stays aware about what is happening underneath the vessel. Standard equipment such as the sonar provide information on fish activity and sea bed structure, and more advanced software enables the display of heat maps, sea bed structure, 3D images and so on.

Also on our long lining trip, the skipper looks at the different monitors to decide where to release the lines most effectively. The same holds true for fishing with jigging computers, in which the skipper has to decide from which spot to start the drift. For instance, the sonar can detect shoals of moving baitfish that give good evidence of active target fish such as cod. Furthermore, maps of seabed structure can, in combination with GPS positioning, help to detect distinct seabed structures such as drop-offs, slopes and banks that typically hold fish. Although certain areas such as the 'fruit basket' just outside the north-western fjords are commonly known among skippers as good jigging-spots, modern navigation technology can help to save time and fuel costs to exactly locate one of the known structures in the area such as the 'banana', the 'apple' or the 'pear'—names given by locals due to their fruity resemblance on the screen.

All in all, our observations from the small boat show how digital information technology enframes the ocean by turning the fish more and more into a synchronised standing-reserve that becomes constantly detectable and retrievable. As a consequence, digital enframing alters the temporal and spatial relations between the fish and the fishers, who's movements and practices are increasingly linked and synchronised with each other. This technological development goes hand in hand with the development of stronger and faster fishing vessels and safety equipment that enable coastal fishers to put to sea in much rougher conditions far-off the traditional coastal fishing grounds within 12 miles from shore. Thus, technological enframing opens up new pathways for practices of fishing by redefining traditional forms of knowledge and skill and literally pushing the boundaries of coastal fisheries into deeper waters (more on this in Chapter 6). In this sense, 'effectiveness' in the first place not only refers to one or another form of cost-reduction, but to the opportunity of revealing new fishing grounds and tracking the movement of fish stocks that bring forth the ocean's potential as standing-reserve. Thus, the role of modern technology cannot simply be understood in terms of increasing rationalisation, but as a way of changing the temporal and spatial horizon of the fisheries towards increasing synchrony with the environment, which allows the opening up and rapid exploration of new fishing grounds beyond the traditional coastal boundaries of the small boat fleet.

Conclusion: The Human Limits to Technology

In recent years, modern technology has created new pathways for practices that have not only changed the relation between the fishers and their environment, but the world of modern fishery economies as such. Part of this transformation can be found in technological development, which has not only translated fish stocks into a manageable resource and commodity, but has also transformed the world of fishing itself. Rather than seeing technology as something separate from culture and society or merely a means–end relation, we have learned that technology itself is world-constitutive and deeply entrenched in the socio-historical context in which it is deployed. The ethnographic material has shown that, in combination with more engine power, digital information technologies today play an important part in the technological enframing of the ocean as standing-reserve. Hence, in combination with scopic media such as chart plotters and computer screens they link and synchronise fishers with the fish and their competitors equally. As a consequence, the spatio-temporal boundaries of the world of coastal fishing are redefined and opened up for new fishing practices that likewise enframe fish and fishers. That this enframing of humans and their more-than-human environments into a techno-scientific management regime does not merely ease the labour required for the efficient extraction of marine resource but also a challenge to human coping, however, becomes obvious on our journey back home.

Although the boat is far from being filled-up after six hours of restless work, the way back to port is much slower with an average speed of 12.6 kn, adding an additional two hours to our sailing time. During this long journey back, it is not seldom a struggle for crew members to avoid dozing off in the wheelhouse due to exhaustion after a long day on deck, especially when the vessel is sailing on autopilot—a dangerous situation when approaching the rugged coastline of the Westfjords! In fact, in the summer of 2013, a brand new vessel from the nearby village of Bolungarvík ran aground on its way back to port as both crew members fell asleep after a long day at sea. Luckily, both crewmembers could be rescued by the Icelandic coastguard's helicopter, though the vessel was a total loss. Besides this dramatic case, this was certainly not the only time a crew or skipper has lost control over a boat due to total exhaustion. Moreover, field observations and interviews suggest that exhaustion and fatigue are a common problem across the fleet. Accordingly, a skipper reported that he felt 'like a zombie' (FN: 96) after fishing throughout the last few days, and two others likewise reported feeling 'really tired' (MIX; FN: 62), but had to go to sea the next morning. During fieldwork, I myself even witnessed a skipper waking up a skipper on another boat who dozed off at the wheelhouse on his way back to port (FN: 97). Apparently, long workdays at sea in a rough and highly intensified socio-technical environment seem to demand their tribute, and exhaustion and fatigue seem to mark the human limits to both skill and modern technology.

Field observations suggest, however, that fishers deploy strategies to cope with physical exhaustion. One of these strategies is, if the skipper is not out alone, based on rotation: after taking a meal, the skipper, who was in charge of monitoring the wheelhouse on the way out, chooses to get some rest in one of the bunks. From now on, the deckhand is in charge of monitoring the vessel during the long journey back to port. Like the skipper on the way out, he will entertain himself most of the time by watching films, browsing the internet, making phone calls, eating, drinking coffee and engaging with all sorts of equipment in order to stay awake and aware of his surroundings.

The watch of the deckhand, however, will end after approximately three hours, as manual navigation remains the untouchable competence of the skipper. Now, the deckhand will wake up the skipper to get ready to manoeuvre the vessel manually into the fjord. The weather is getting better and the boat starts cutting more and more smoothly through the sea the closer we get back to the coastal region. Shortly before we arrive, the skipper will fill out his (analogue) logbook with information on date, catch and position. Just before we arrive, he will call the coastguard to report that the boat is about to return to port.

Back at port, however, the workday is not over yet. The catch, which is stored in containers of approximately 250 kilos each, has to be unloaded manually with the help of a crane at the harbour and will be picked up by a forklift that will bring it either to a local processing plant or to the local fish auction; the fish boxes and the deck are hosed with seawater and the wheelhouse is cleaned. Finally, after almost 16 hours at sea, I am finally off duty. For the crew, however, the rest is short, as the next fishing trip is already scheduled for the upcoming night.

But why do fishers fish themselves to exhaustion if technology is providing them with ever-more efficient capture technologies, complex navigation software and ever-faster vessels? The Heideggerian analysis of technology stops short with regard to this question, as it does not take into account the structural conditions and global entanglements of market-based fisheries. When understanding the economy as a more or less autonomous domain of coordination, however, issues of risk and uncertainty come to the centre of attention in understanding the dynamics of modern market societies. The next chapter will attempt to map out the flows and tensions between modern technology and human coping in contemporary rural capitalism.

References

Callon, M. (1998). An Essay on Framing and Overflowing. In M. Callon (Ed.), *The Laws of the Markets* (pp. 244–269). Oxford: Blackwell.

Castells, M. (2001). *The Internet Galaxy: Reflections on the Internet, Business, and Society*. Oxford: Oxford University Press.

Haraway, D. J. (1985). A Manifesto for Cyborgs: Science, Technology and Socialist Feminism in the 1980s. *Socialist Review, 80*, 65–107.

Heidegger, M. (1977). The Question Concerning Technology. In M. Heidegger (Ed.), *The Question Concerning Technology* (pp. 3–35). New York City: Harper Collins.

Holm, P. (2001). *The Invisible Revolution: The Construction of Institutional Change in the Fisheries* (Ph.D. dissertation). University of Tromsø, Tromsø.

Holm, P., & Nolde Nielsen, K. (2007). Framing Fish, Making Markets: The Construction of Individual Transferable Quotas. In M. Callon, Y. Millo, & F. Muniesa (Eds.), *Market Devices* (Vol. 55, pp. 173–195). Malden: Blackwell

Johnsen, J. P., Holm, P., Sinclair, P., & Bavington, D. (2009). The Cyborgization of the Fisheries: On Attempts to Make Fisheries Management Possible. *Maritime Studies, 7*(2), 9–34.

Knorr Cetina, K. (2003). From Pipes to Scopes: The Flow Architecture of Financial Markets. *Distinktion: Scandinavian Journal of Social Theory, 4*(2), 7–23.

MacKenzie, D. (1984). Marx and the Machine. *Technology and Culture, 25*(3), 473–502.

MacKenzie, D. (2009). *Material Markets*. Oxford: Oxford University Press.

MacKenzie, D., & Wajcman, J. (1999a). Introductory Essay: The Social Shaping of Technology. In D. MacKenzie & J. Wajcman (Eds.), *The Social Shaping of Technology*. Buckingham: Open University Press.

MacKenzie, D., & Wajcman, J. (Eds.). (1999b). *The Social Shaping of Technology* (2nd ed.). Buckingham: Open University Press.

6

When the Fish Ignore the Market

This fucking quota system, it's unbelievable!
Skipper on the 'haddock-crisis'

When abundant resources become a problem…
Summer has finally arrived at the Westfjords. June has been tremendously warm with temperatures of up to 17° and unusually
calm seas—a relief for small-boat fishers such as Bjartur around the
Westfjords region who spent most of their wintertime tied to the docks
due to long periods of relentless storms. Although fishing on small boats
has widely become a year-round, full-time profession, usually the largest
part of the quota is caught during the winter season when the fish are
moving slowly and are willing to take a lazy bite on some fresh oily bait.
Due to the harsh winter, however, many were not able to fish as much
as they would have liked to. As the fishing season will end with the last
day of August, many now start feeling the pressure to finally finish off
the season's quota share.

It seems that the most rational way to catch a lot of fish in a short
time would be to gear up for a lot of longlining trips to make sure to
cash in as rapidly as possible. Fishing in the summertime with 24 hours
of daylight for most months, however, does not necessarily mean that

© The Author(s) 2019, corrected publication 2020
A. Dobeson, *Revaluing Coastal Fisheries*,
https://doi.org/10.1007/978-3-030-05087-0_6

fishing on calm seas is more convenient, as other forces than weather have troubled the coastal fishers of the Westfjords region in recent years. Nevertheless, it is certainly not the lack of fish outside the fjords that is the problem. In fact, rather the opposite is the case as the Westfjords have had the reputation among Icelanders of having short access to some of the most productive and stable fishing grounds ever since the days of settlement. What, then, is troubling Bjartur and the other small-boat fishers?

Icelandic folk mythology tells the tale that when the Devil tried to catch a haddock, it was so slimy that it slipped out of his hands, leaving a black mark of the Devil's thumb by the gills, followed by a long black line over the lateral organ. And today, it might seem to the superstitious that the Devil has cast a spell on the haddock: what used to be a welcome by-catch and traditional basis for the production of traditional dry fish (*harðfiskur*) in Iceland and a popular ingredient for fish 'n' chips in the United Kingdom, has become more and more troublesome on the lines of the small-boat fishers around the Westfjords.

Despite the daunting prognoses of the Marine Research Institute on the development of the haddock stocks,[1] which has impelled policymakers to cut the haddock quota in recent years, many fishers, paradoxically, have become troubled by the rich fishing grounds just outside their fjords, where the highly priced haddock seem to congregate despite the predictions.[2] When the quota system was implemented, however, haddock were not really an issue in the area and most fishers chose to invest primarily in their cod quota instead, which has formed the backbone of the industry for decades. As a consequence, many quota-owners are not among the lucky ones who invested in a lot of haddock quota and today are struggling with the paradoxical situation of catching too many fish, as a skipper explains:

---

[1]The Total Allowable Catch for haddock has seen a steep and steady decline from 93,765 tonnes in the fishing year 2007/2008 to a mere 27,404 tonnes in 2014/2015 (cf. Fiskistofa, 2015).

[2]The abundance of a species around a certain region does not necessarily contradict the results of stock assessment, as fish are known to congregate in a few areas in times of declining stocks (interview with marine biologist, Copenhagen, 2011).

AD: I heard that people sometimes want to catch cod and then they get too much haddock?

Skipper: Yeah, that's a big problem now everywhere, for the last two years! They are always putting down the quota on haddock, but there is haddock all over. But it seems that people who are making the quota, they don't want to listen to the fishermen, it's strange, they never listen to people who know most about it, that's my experience! (MIV)

Not only does the skipper point to the typical conflict between scientific and local knowledge that characterises modern resource management regimes all over the world[3]; he also points out an economic problem for many fishers that is the result of a mismatch between the political construction of scarcity and the actual environmental conditions that confront many quota-owners and fishers from the Westfjords region:

We are supposed to bring the whole catch ashore and have to match the amount of fish we bring ashore with allotted, leased or bought quota. Haddock is a well-known problem, a lot more is caught than your quota can cover, so you have to lease or buy to balance the difference. (MXIV)

Although the quota market generally allows free transfers and leases of fishing quotas from boat to boat, the haddock crisis has heated up the lease market too much to make it a long-term solution, as Bjartur who's family has been severely constrained limited by this problem explains:

AD: So it's expensive to lease quota for haddock?

Bjartur: Yeah, it's soo expensive!

AD: And do you sometimes have to if you get too much?

Bjartur: We try not to, because it costs you 315 crowns to rent the haddock—and listen carefully now—we are getting 270 on the market! You have to rent it in a higher price than you are getting fish on the market! (…) You are losing money in every kilo you are fishing!

---

[3]For an example from North America, see Acheson (2003: 145).

Bjartur's calculation makes clear that, no matter how one puts it, the haddock has been transformed from being a welcome and highly valuable by-catch to a costly loss in the accounts of many fishers and quota-owners from the Westfjords, as additional quota must be leased from bigger companies who can afford the profitable business of renting out some of their quota shares at sky-high prices (see Appendix, Fig. 6). Hence, due to the dire need of many quota-owners to rent haddock quota, the lease price sometimes even exceeds the price for raw materials, leading to severe losses in the accounts of many small-boat fishers if the haddock is in the mood for a feast.[4]

In the Westfjords region, these different dynamics today have led to the paradoxical situation of fishers with too many fish—'the wrong fish' due to the lack of fishing quotas for haddock in their accounts. In fact, the problem of landing haddock has started to create serious problems for some small-boat owners, who not only need to cover the costs of baiting the lines and the sailors, but are facing financial pressure due to their financial entanglements. Moreover, also fishing crews' wages depend on the total value of a vessel's catch, which makes them share a common interest in avoiding the landing of haddock. With the pressure from their creditors breathing down their necks, falling profits have become a serious issue especially for small-boat owners around Bjartur's homeport community where the haddock are plentiful.

It becomes apparent that that the fish do not seem to care much about the boundaries of the quota market, and the recent 'mackerel-wars'[5] about 'British' and 'Irish' mackerel moving north to the colder waters of Iceland and the Faroe Islands indicates that this is anything but an individual case of marine inhabitants ignoring the territorial boundaries of property rights and the law of the sea. But our small-boat fishers could simply not foresee the haddock moving into their fishing grounds when investing the cod quotas. But what can they do?

---

[4]Auction prices fluctuate a lot, and prices can be well above the lease price, which usually occurs during storms or holidays—times when coastal fishers usually stay ashore. When the prices are low—around 300 ISK—the rent price is likely to be on a par with the lease price of around 300 ISK, leaving no surplus to the coastal fishers (based on RSF, 2015).

[5]For the case of the 'mackerel-wars', see Bazilchuk (2010).

More-Than-Human Market Environments

It has been rightly criticised that it is not enough to endlessly point out that the rational actor model is empirically wrong (Beckert, 1996, 2009). For this reason, economic sociologists have revitalised Frank Knight's (1921: 101–120) seminal distinction between incalculable *uncertainty* and calculable *risks* as fundamental concepts of New Economic Sociology (Aspers, 2018; Beckert, 1996; Dequech, 2003). Accordingly, although actors can rationally calculate the risks of economic transactions, the future can never be known and remains uncertain. Thus, it is this fundamental uncertainty that forms the driving force of the economy, rather than a state of equilibrium between supply and demand, as suggested by neoclassical economics. Thus, instead of simply assuming some form of 'bounded rationality', sociologist studying the economy should rather look at the 'cognitive, structural and cultural mechanisms that agents rely upon when determining their actions without knowing what to do in order to maximise their outcome' (Beckert, 1996: 814–815).

Although this approach has created valuable insights into the social, cultural and institutional aspects of valuation and exchange that drive economic decision-making in concrete empirical markets,[6] this sociological view on the economy nevertheless remains limited to the cognitive limitations of the isolated intentional actor, which we have questioned in Chapter 4. Hence, economic coping cannot be understood as the mere result of intentional framings of a situation, nor to the blind rule-following of a distinct set of habits and cultural scripts. Rather, economic practices are embedded in a social-technical context in which skilful routinised practices allow flexible adaption to the situational contingencies of an ever-changing world. Moreover, especially in natural resource-based economies such as fisheries and farming, economic actors must not only cope with the problem of decision-making

---

[6]Instead of claiming the existence of one general market model, the sociology of markets rather points out empirical variations across different types of markets. For a comprehensive review of the field, see Aspers (2011).

'in' concrete markets, but with multiple sources of uncertainty stemming from their entanglements with the socio-material and ecological contingencies of a 'more-than-human' world.[7]

Especially the haddock crisis makes clear that in market-based fisheries, skippers and quota-owners are confronted not only with the unknown futures of investments and exchange in markets, but with multiple, in particular institutional, technological and ecological uncertainties stemming from the broader political, techno-scientific and ecological entanglements of market-based resource management regimes. While especially the economy, politics and law can be influenced, depending on individual resources, natural phenomena such as weather and migration of species seems largely out of human control. Nevertheless, these sources of uncertainty can be translated into calculable economic risks that allow for a certain degree of predictability. It is for the purpose of risk-reduction that science and of science and technology play a special role in modern market economies (Luhmann, 1993: 95).[8] This becomes in particular apparent in natural resource-based economies, which are entangled into the flows of globally linked eco-systems. In order to allow for the sustainable exploitation of marine resources, marine scientists have therefore developed astonishing tools such as Virtual Population Analysis (VPA)[9] for assessing, predicting and managing the development of fish stocks. This technological enframing of the sea reduces the complexity of the marine environment by breaking it down by attribution of simple causalities.[10] In other words, scientists must simplify 'nature' to isolated parameters in order to translate the highly complex nature of marine eco-systems

---

[7]The notion 'more-than-human' was originally introduced in the fields of cultural and human geography in order to correct the anthropocentric bias of previous perspectives by highlighting the socio-material and environmental aspects of space (Braun, 2005; Whatmore, 2006).

[8]According to Luhmann (1993: 93), the more society relies on money as generalised means of communication, science and technology become important means for risk reduction.

[9]For a brilliant sociological account of the role of VPA in the construction of markets in the fisheries, see Holm and Nolde Nielsen (2007).

[10]Or in the jargon Luhmann's system theory: technology is a *functioning simplification in the medium of causality* (italics in original) (Luhmann, 1993: 87).

into calculable arrangements that can be managed and controlled. The underlying models and predictions, however, are fed empirical data based on standardised samples taken over the years and represent only a fraction of a globally cross-linked and highly complex and ever-changing eco-system. What remains bracketed in these attempts to control the environment are thus the complex arrangements and local contingencies of a globally interlinked and hypercomplex marine eco-system. In the long run, the 'fluid spaces of the sea' (Bear & Eden, 2008) thus tend to provoke repercussions by undermining the territorial boundaries of techno-scientific resource management regimes, inevitably giving way to new sources of uncertainty. The higher the financial stakes are in the industry, the more scientists will have an incentive to optimise their models and predictions to match the flows of the sea. In this sense, 'problems of technology reveal themselves in the paradoxical attempt to solve the problems of technology by technical means', as Luhmann (1993: 90) puts it. However, these new models and predictions are nevertheless prone to be undermined by a highly volatile and ever-changing world, spurring a money-induced cycle of socio-technical problem solving and breakdowns.

The example shows that attempts at risk reduction bracket the fact that the complex spatial flows of the sea cannot be fully controlled: one can construct the perfect market, but what can one do if the fish don't know about it? But while scientists, economists and engineers continue chasing the modern dream of total control over nature, our small-boat fishers are in a more immediate confrontation with the flaws of modern resource management. But are they simply doomed to accept this mismatch between the 'laws of the market' and the 'forces of nature'? To answer this question, we will continue following Bjartur and the other small-boat fishers on their daily grind and further decentralise the conventional actor-centric view focusing on exchange in markets to the socio-technical practices economic actors deploy in a highly volatile and ever-changing more-than-human market environment.

Adjusting the Market to the Flows of the Sea

Bjartur knows that his current financial hardships are not the haddock's fault. But while small-boat fishers have a hard time influencing the predictions of marine biologist, they also have little power to influence the politics of the quota-market in a substantial way. So if the facts don't fit the theory, why not try making them fit? But not by changing the assumptions of the theory, but by manipulating the facts to fit into the techno-scientific and bureaucratic world of market-based resource management!

From the perspective of the fisher, the haddock-crisis is first and mostly a practical challenge akin to our technological breakdowns from Chapter 4: the more haddock find their way into the hold, the more they are prone to be problematised as economic loss. Thus, simply throwing unwelcome fish back into the sea before returning back to port seems like the most effective and cost-effective solution that is widely practised in many quota-based fisheries around the world. The calculation is simple: if the fish is not landed, it can't be deducted from the vessel's quota share. Unlike in many other fisheries, however, the practice of discarding is strictly prohibited in Icelandic waters. Hence, if fishers are caught discarding their fish by the coast guard, they risk costly lawsuits, severe fines and potentially even the loss of their fishing licence—in the long run just too high of a risk in today's panoptic fisheries of mutual surveillance and control (see Chapter 8).[11] But what else can fishers do to legally cope with this difficult situation?

In despair about the wrong fish in their fishing waters, Bjartur and the other small-boat fishers have started to skilfully manipulate their socio-technical environment in order to adjust the market with the reality of their fishing grounds. As a result from finding ways to deal with excess landings, three analytically distinct coping practices have emerged from the every-day coping with the haddock crisis: (a) tinkering with accounts; (b) socio-technical conversion; and (c) redefining boundaries.

---

[11]This obviously does not mean that discarding is not practised at all in the Icelandic fisheries. Still, the high-stakes and surveillance makes it very risky (see Chapter 8).

Tinkering with Accounts

From a short-term economic point of view, it seems rational to invest in more haddock quota for the next fishing season when the fish abound in a region. The reality of small-boat fishers, however, looks different. Often already carrying the burden of substantial debt in their accounts, investing in more quotas while at the same time seeing a steady reduction of the Total Allowable Catches (TAC) from the government seems to be far too costly a solution, which only big companies can afford. Instead, quota-owners rather look for short-term solutions over the season in order to keep their operations running. I call this coping strategy *tinkering with accounts.*

In contrast to long-term planning strategies, tinkering with accounts is a way of matching a company's accounts by engaging in short-term market-transactions to cover excess catches. The most basic practice of tinkering with accounts is renting haddock quota on the lease market. This allows a small-boat to continue fishing operations despite substantial by-catch of haddock. These transactions, however, are not profitable as such, as the lease price tends to exceed the market price for raw materials. Hence, leasing quota is an economic trade-off between potential losses from the lease price and the income provided by the core business of a company, which usually is the cod fishery. Bjartur explains this strategy:

> Ja, some guys, you know, on the longlines, they have to have haddock quota because you cannot say (to) the fish: I just want the cod on the longline, no haddock, you can leave! – You just cannot do that. What bites on the hooks comes up and you have to have quota for it, and they are renting the quota in haddock so they can fish the cod, that is the only reason they are renting haddock. So if they don't have the haddock, they cannot fish the cod they have, you know what I mean? (MII)

In this case, quota-owners renting haddock quota for a higher price than they can sell the raw material is seen as a necessary cost that allows fishers and quota-owners to keep up with their daily business of catching cod. Quota owners therefore must act quick in order to cover the excess catch in order to avoid hefty fines, as Krístin who manages the accounts of Bjartur's company explains:

> I never had a problem because I have a very good friend who is in quota selling in Reykjavík, I just call her and say: I need this today, uh, you have to be quick! And I always fix it, I never have a problem! (MVIII)

Because there is no official market platform on which all lease offers are collected and information is distributed, leasing quota is highly dependent on network ties and brokers that mediate between the different parties. Hence, the contact with the broker in Reykjavík gives her the advantage of fulfilling the legal requirements on time. When asking whether her connections in Reykjavik put her in a better bargaining position for prices, however, Krístin also makes clear that this is not the case. In the long-term this means that if catches of haddock maintain over a long period of time, the overall profit of the company is reduced. Moreover, the closer the fishing year comes to its end in August, quota leases become more costly, as larger shares have already been fished up during the season. Often, vessels have already exceeded their allotted quota and are running the risk of costly fines for the owner by the end of the fishing season. Especially for small companies trying to finish their regular cod quota but have caught excessive haddock this can be a problem because they lack the financial means to rent large amounts of increasingly expensive fish. In this case, quota-owners with strong ties that trust each other may help out tinkering with each other's accounts to avoid costly fees, as Krístin explains:

> And sometimes if you have a problem by the end of the fishing year, if you know someone who trusts you, he can let you have quota and you can give it back, you know … If some fisherman wants quota, they know they can have it after few days, I can put the quota on their boat and they can put it back and the same if I need quota and someone can help me to let me have quota on my boat and they have to write the papers, and you have to fax it with fax – I always can fix it, it's no problem, not for me. (ibid.)

When leasing haddock, tinkering with quota accounts may also include renting out fishing quotas that are of little use-value for the quota owner. In this case, 'market devices' (Muniesa, Millo, & Callon, 2007) such as the 'quota calculator' (see Image 4.5) become an important means that enable calculability as basis for economic tinkering:

> Sometimes I need haddock, and then I put another kind, what you call it, saithe, we cannot fish that fish so much and then I rent it away and take another kind into my boat and then I have to look into my computer and see how much, and in end of August [final month of fishing season] I do this. (ibid.)

It becomes clear that the economisation of the small-boaat fisheries has created a system that requires quota-owners not only to engage in rational long-term planning for a fishing season, but to engage in short-term transactions, networking and market observations to adjust their accounts with the highly volatile and ever-changing environment of the sea. Although Krístin's confident remarks suggests that her network ties help her to fix any quota-related problem, it is obvious that the company has an interest in avoiding these type of transactions, which nevertheless involve a high level of uncertainty and economic losses. Furthermore, the more the annual haddock quota is reduced by policy-makers and the more the demand on the lease market increases due to increasing catches, even the best network ties will not help a company from running into high extra costs if a vessel lands a lot of haddock. Instead of asking for new loans to invest in expensive haddock quota, fishers have developed strategies of avoidance to reduce their annual haddock landings. Two of the most common practices will be presented in the following.

Socio-Technical Conversion

A common strategy of cost-reduction is socio-technical conversion, which in this context refers to a coping strategy that allows fishers to control costs and landings by swiftly adapting capture techniques to changing conditions at sea. In this case, socio-technical conversion is a practice that allows fishers to switch between different types of fishing gear depending on the weather situation, type of fish and fishing season.

Fishing with long lines is considered by far the most effective capture technique over the winter months when the metabolic system of the fish is rather low. On average, a fisher expects 700–800 kg fish per fishing line. When the water heats up over the summer months, the catch can fall below 100 kg per line, which is not considered profitable anymore (FN: 41). For this reason, fishing with 'active' bait by means of jigging computers has proven to be more cost-effective in terms of catches over the summer months, when the cod is chasing after huge shoals of bait-fish. Struggling to pay his bills in the post-crash era, jigging has also become a very attractive alternative for Bjartur to lower the costs of labour, fuel and bait:

> I have a lot of quota in cod and now I've started to choose computers because I don't need two persons to baiting a long line, I can be on one boat, just myself on the boat alone and I can make a good process, you know, for me myself and the company, because I don't have to pay as much salary with the computers. And last summer we worked with the computers and we had 60% or 50% of the money that was left in the company, but with the long line, we are sometimes going down to 10% left of the money into the company, which makes it [difficult], when it's so low, so low, just 10%. You do not manage to pay the credit, to buy a new line, to buy the oil, to pay the new tax, which the government has put on all the fishing! (MII)

Due to the legal statutes of the hook-and-line quota system, small boats under 15 (now 30) tonnes can choose freely between either longlining or jigging (see Chapter 3 and Appendix). In contrast to a longlining operation, the boat drifts when using the jigging computers and the artificial rubber baits can usually be used for many trips until they are lost or worn. Due to the high operating costs of longlining and its lower effectiveness in the summer months, many small-boat fishers therefore deploy hybrid vessels, which allow them to switch between different

Image 6.1 The hybrid fishing vessel. A typical smaller hybrid vessel (8.44 BT) in the winter waiting at port for better weather to go out longlining. Over the summer months (June, July and August), the skipper will install jigging computers on the port side of the vessel (see Image 6.2) (Photo by AD)

types of fishing gear, depending on fishing season and weather situation, as jigging itself requires a fairly calm sea state to allow the smooth and slow drift that is key to success (see Images 6.1 and 6.2).

Lately, however, socio-technical conversion has not only become an important way to reduce overhead costs, but a deliberate strategy of coping with the overabundance of haddock in the region. The reason for this is very simple, as Bjartur reveals: 'We want to fish with the computers because we just get the cod on the computers, there is no haddock…' (MII). Although he shows little interest in my theories of why the haddock are not taking the bait—'It is like that' and 'fact is that the cod is taking on the computers' (MII)—he tells me that the family is 'thinking to change the boat over to computers, maybe for 6–7 months over the year and have the long line for the rest, yeah, for 5 months'. In fact, it will turn out that Bjartur is not alone in his knowledge of the haddock's behaviour, because other vessels from the region are following

Image 6.2 Summer jigging. The same vessel as shown in Image 8.3 equipped with jigging computers for the summer season. Although the overall operating costs are lower than for longlining, a jigging trip lasts much longer than a longlining trip (up to 48 hours) to land an average catch of only 2–3 tonnes (MVII) (Photo by AD)

a similar strategy—even if their vessels are not really suitable for conversion. For instance, when talking to another skipper, vessel at the docks, I was wondering why even one of the newer 15 tonne longliners—similar to the one illustrated in Chapter 4—is equipped with jigging computers (FN: 97). Again, the answer is the wrong fish: 'They don't have so much quota, and they do it because of the quota, their haddock quota is very low' (ibid.). Due to its wide hull and half-deck, however, the skipper has some difficulty controlling the vessel for a good drift as it is very sensitive to wind due to its size, or in the words of the skipper: 'It is like a sail' (ibid.).

Although this coping strategy can be seen as a clever and 'creative' way of swiftly adjusting the territorial boundaries of market-based fisheries to the local contingencies of the sea, they—like many technical solutions- inevitably produce sociotechnical drawbacks in other areas. In this case it is not mainly the small-boat fishers themselves, the vessel or the fish, but the community which benefits from the labour- and cost intensive longline fisheries. Far from being opportunistic, especially fishers from struggling communities such as Bjartur are aware of this dilemma:

> That makes it difficult because you have people who is working for you and the what are you gonna do with the people, ja. So that makes it very difficult, and they cannot wait for seven months and come again when I need them, so it's confusing, you know, it's not working together … It's okay to take maybe two, three months and we say 'You take off free one month' but 7 months, 6 months is too long. They need to have the money, especially now [after the financial crisis], it's a tough time and everybody needs to have money in every month!

While it is clear there is a tension between perceived economic pressures that push quota-owners towards increasing cost reductions, on one hand, and social responsibility towards the members of the community on the other hand, socio-technical conversion itself seems to be an attempt to balance cost-effectiveness with community responsibility by having 'both jigger and line in the summer (…), they go maybe once or twice a week with the line so the baiter will have some work, then he is jigging' (MIX).

At the time of writing, it was unclear to what extent this goodwill could withstand the pressure towards increasing economisation. Just before I left the community in 2014, however, Bjartur's family dared to invest in a brand-new vessel despite their financial hardships. In contrast to the family's previous, smaller vessel, which certainly showed signs of heavy usage over the years, the new vessel appeared to be not only bright and shiny when sailing over from the wharf, but also slightly larger (11.78 BT) and more than twice as fast as the old vessel (27 kn compared with 12 kn). Of course, it also outclasses the old vessel in terms of navigation technologies and observational scopes. Most interestingly, however, was a special device underneath the wheelhouse by the bunks, which was proudly presented to me at the launching:

a drift bag, which can be launched manually when seas turn rougher to maintain a smooth and slow drift to preserve perfect conditions for jig-fishing (FN: 82–83). When returning to the community in 2018, however, the pressure to economise seemed to have prevailed and Krístin reveals that they have completely stopped longlining over the summer in order to keep up with their payments: 'It's sad, but the guys from the baiting house now have to live on benefits'.

Redefining Boundaries

A third strategy for reducing the landings of haddock is simply fishing where it does not abound: in high seas. For this reason, I call the third strategy for reducing landings of haddock *redefining the boundaries* of the small-boat fisheries, which describes the practice of moving the traditional coastal boundaries of the small-boat fleet further into the open waters of the Arctic Ocean (see Image 6.3). As a skipper makes clear:

> We have to go very long to get cod … Today the haddock … I don't know how … He is everywhere! So it is very difficult if you are just fishing cod. I think maybe in the summer it's not a big problem for us to go to sea, [but now] maybe you have to go like 30–40 miles to just go over the haddock. (MI)

In fact, it would even turn out that the reason for going out a long way offshore in Chapter 4 was not merely the promise of a good catch, but a strategy for avoiding catching haddock. But while fishing offshore has become a widespread practice in some regions for these reasons, it can only be practiced by those who have 'good boats (…) and the quota to do it', as another skipper who attributes a certain degree of 'craziness' to these fishers puts it (FN: 63).

Sailing offshore on small-boats, however, not only involves physical danger, but also additional economic risks: the crew must land more fish than when fishing closer by in coastal waters to compensate for extra fuel costs. As it will turn out, this is not always the case despite

Image 6.3 Redefining boundaries. The upper group of vessels (Hrolfur Einarsson, Frida Dagmar, Sirry IS-84, Tjold, Tryggvi Edvards) north off the Strandir region (the northernmost tip) to the north shows a pattern of newer 'small' longliners congregating about 50–60 nm offshore in June 2013. Below, there is another group of vessels fishing fairly far out (Smari IS-144, Steinunn, Hjortur Stapi). In contrast, one can also clearly see some traditional small-boats vessels fishing very close into the coastline

modern technologies such as fish finders on board: during our fishing trip in Chapter 4, the previous motivation and euphoria of the crew already faded after the first line was hauled up: instead of heavy lines, the skipper was only gaffing some odd smaller fish, and his experience was telling him that he had laid the lines in the wrong spot, although a friend of his reported a really good haul just one day before: 'It is always a risk you take when you go that far out. Either you hit the jackpot or get nothing! (FN: 77)' To make the extra costs, labour and time spent on the high seas worth the journey, the crew had to literally 'fill up the boat' with 10–15 tonnes of fish. Instead, we only landed around 5 tonnes, which would be a good catch in coastal waters but way below the expectations in this case. But not only high economic risks are at play with this strategy.

While sailing back to port I ask the crew if they have ever been in trouble when fishing far out offshore. The deckhand just grins and looks at the skipper: 'Tell him about it!' The skipper tells me that he once ended up being in distress at sea when fishing a similar spot. On that particular day about two years previously, big winds turned up and a wave of about three metres hit the boat so that it almost capsized and half of the bala and fish went overboard. After putting on the survival suits and requesting the coastguard helicopter in case of disaster, it was only because of the skipper's skill and the nearby trawlers shielding the vessel from being turned upside down by the growing waves that the vessel made it back to port. When asking the same skipper if he did not foresee that a storm would be coming up, he explains:

> Yes I knew that it was coming, but it was nice weather like this and I thought I could get away with it. But see, it also has something to do with this quota system: we've been catching a lot of haddock that time closer to shore, but the haddock quota on this boat was very low. So I had to take some risk and go far out to get more cod. (FN: 80)

Obviously, fishing always takes place in a potentially rough and dangerous environment. But this example makes clear that physical danger and economic risk lie very close to each other, as the margins of sea-worthiness for a smaller vessel are very slim in the unpredictable environment of the high seas, and even the latest navigation technology providing underwater maps is no insurance against potential economic losses. Nevertheless, it has become a widely-practiced strategy to take these risks in the hope of a good cod-haul, especially for the newer, bigger longliners in the small-boat fleet who today more and more share the same fishing grounds with the large trawler fleet.

Processing Uncertainty

While the examples above have focused on the problems of fishers short of haddock quota, some quota-owners specialised and invested in systematically in lots of haddock quota before the quota was cut by the government. That this does not necessarily protect these companies from the hardships of the haddock-crisis will be demonstrated in the following, as an example from one of Bjartur's neighbouring communities in another fjord shows.

Putting everything on one card

In contrast to the bulk of investors, the remaining members of a small fishing community in decline saw the potential of specialising in haddock due to the good fishing grounds just outside the fjords. For this reason, they started to rebuild their business collectively in the early 1990s based on small-boats like many other fishing communities have tried in the Westfjords. In contrast to the fishers from the villages above, the fishing grounds close to the company's fjord were known among locals as good and productive haddock spots, which led the local processing plant and the fishers from the community to specialise in haddock, as the CEO of the local processing plant recapitulates:

> So we actually decided on haddock, there is always a lot of haddock in this area, we try to focus on haddock. We put everything on one card. (MXIX)

With the emergence of the fish auctions and the shift in global markets, the company created a market niche and invested in more and more haddock quota over the years after a number of ups and downs and merging with a bigger company in 1997. The merger allowed the company to invest in more haddock quota to expand the company's business, and the strategy of specialisation in haddock was paying off with large contractors in the United States and the United Kingdom. With the general decline of the Total Allowable Catch (TAC) for haddock, however, the company's vessels were forced to land fewer fish, as one skipper explains:

> *Skipper*: We are specialist (sic!) in fishing haddock, mainly building a market for haddock, but ah, they have cut down the quota so much for haddock…
> *AD*: Yeah, the price is really high…
> *Skipper*: Extremely high! So it's difficult to grow, because we need more quota in haddock, and maybe the price is getting higher and higher for haddock, so we are trying to fish haddock, but it's not really much fishing, maybe one and a half tonnes per boat, per day! (MIX)

This, of course, gives the fishers less work and consequently reduces the amount of raw materials for processing, according to one skipper fishing for the company by more than 50% (MIX). To cope with this problem, the company is forced to buy more and more raw materials at the fish auction to keep the factory running and fulfil the contracts with their buyers. As the overall TAC for haddock was lowered over the years, however, the general supply at the fish auction decreased and raised prices. After all, the company is facing the luxury problem of specialising in a product that has become too valuable for the international export market.

In general, processors are highly vulnerable to the fluctuations in catches, as they require raw materials in order to keep their production running in order to meet contracts with domestic and international buyers and maintain employment for the members of the community. In particular, processors relying on small-boats are highly vulnerable to the weather situation in contrast to processors relying on large trawlers that are more sturdy and can easily be moved to the other side of the island in hope of calmer seas. As a consequence, the haddock crisis directly impacts the international business relations of the processing industry.

International buyers are obviously aware of these strong price fluctuations, as the CEO of a medium-sized processing plants from the Westfjords region explains:

> I can tell you about our haddock loins: the usual price is 8.4 British pounds per kilo. The price for last month has been around 12–13 pounds because of the price on the auction. And it has gone once up to 17 pounds per kilo. So they are flexible, they raise the prices if they need,

they trust us to decide it. And of course all the information are (sic!) on the internet, they can just see the prices on the auction. (MXIX)

While the quotation suggests a rather high degree of acceptance of buyers with regard to price fluctuations, however, it will soon become clear that there are limits to what buyers will accept. In fact, while waiting at the office, the CEO will receive a phone call from a broker. After a few minutes of tense conversation on the phone, he will look at me in shock and say:

I was waiting for that call. Now the prices for haddock are so high that they won't take it anymore. (FN: 62)

The example makes clear that rather than struggling due to the lack of fish in the sea or a lack of demand for their produce, the company is facing a high degree of uncertainty with regard to scientific stock assessments, international currency markets and environmental conditions. It seems as the only options left for the company are keeping the fish in the 'bank', as one employee calls the freezing house until prices stabilise again, or selling it at a lower price, which means a short-term loss. These two options of course bear the problem that the future development of prices, government regulations and supply of raw materials, which all impact the market price, are unknown.

While the company cannot directly influence international market movements and government decisions, it had to find alternative strategies to compensate the losses of the haddock crisis. One of them is reducing overhead costs, for example, by buying cheaper bait and not investing in newer technologies for the boats, stagnating salaries and even selling-off company shares and fishing quotas to maintain financial liquidity. These practices, however, are detrimental to the fishers, who suffer from outdated fishing vessels and generally lower catches and salaries. Again, the fishers at sea seem to be the weakest link in the value chain who are pushed to finding new adaptive strategies in order to stay afloat in an ever-changing and unpredictable market environment.

Table 6.1 Coping with the haddock-crisis

Coping with the haddock-crisis	Practice	Sources of uncertainty	Resultant problems
Tinkering with accounts	Matching of accounts with landings, regulations and markets	Fluctuating market prices- and changing regulations	High costs
Socio-technical conversion	Cost-reduction by fishing method (selectivity, fuel cost)	Weather, sea state, fish activity	Conflict with baiters, Vessel might not be suitable
Redefining boundaries	Selectivity, chance of exceptional haul	Weather sea state, fish activity; economic risk and danger	Attrition of engine

Concluding Remarks: The Limits of Market-Based Resource Management

We have now seen that the haddock's unwillingness to cooperate with the predictions and regulations of resource economists, marine biologists and politicians has forced Bjartur and his friends to find ways to making their marine environment fit with the laws of the market by skilfully manipulating and adjusting their socio-technical world. Doing so allowed us to identify three different coping practices: *tinkering with accounts, socio-technical conversion* and *redefining boundaries* (see Table 6.1). The examples show how daily skilful coping at sea responds to the tensions created by the mismatch between the inert territorial boundaries market-based resource management and the highly volatile and fluid spaces of the sea. All three strategies include a distinct technique: as the name indicates, tinkering with accounts is based on the skilful matching of a company's account with the landings of the company and the rules and regulations dictated by the government. Socio-technical conversion includes a modification of the fishing operation by means of switching between alternative equipment that adapts to the environmental conditions and allows for selective fishing to reduce overhead costs. Redefining boundaries solves the problem of spatial congregation by 'sailing over' the stocks, which leads coastal vessels far out onto the high seas.

We have already witnessed, however, that any practice aimed at resolving the tension between the market and the sea somehow seems to produce a new set of problems in the real world. Although all three coping practices are swift and skilful responses to adjusting the highly volatile environment of the sea to the market, they paradoxically confront fishers with new sources of risk, uncertainty and even physical danger: tinkering with accounts is based on the uncertainty of changing rules and regulations and volatile market prices; socio-technical conversion is highly dependent on resources and materiality, which determine feasibility and success; redefining boundaries comes with a high economic risk of having to compensate for the extra costs of the fishing operation and an increasing danger with regard to sailing offshore. Furthermore, coping practices not only involve higher risks, but also come with a number of long-term problems: tinkering with accounts is usually associated with high costs due to scarce resources and sky-rocketing lease prices, which can become a problem for a company in the long-run with regard to mortgages and salaries to be paid. Socio-technical conversion bears the problem of creating a social vacuum, as baiters, as members of the community, cannot only be employed during the longlining seasons over the winter months. Furthermore, as the examples show, some vessels—in particular, bigger long liners with wide hulls—might not be suitable for jigging after all and increase pressure and stress on fishing crews. And redefining boundaries bears the danger of serious technical breakdowns, in particular of the engine due to attrition[12] and to rapidly changing weather conditions out on the Arctic Ocean. In addition to this, long trips bear an economic risk that cannot always be accounted for with regard to the uncertainty of a vessel's catch.

---

[12]I witnessed engine failures on numerous occasions during my time in the field, in particular of larger vessels known to be exposed to a lot of pressure to go to sea and fish offshore over the summer months. This problem even led to a fishing trip I was invited on being cancelled, as the vessel's engine broke down completely so the crew had to wait more than a month for the new parts to arrive (FN: 51; also FN: 63, 70; MXVI).

It has now become obvious to us that technology does not only provides solutions to problems and breakdowns, as its creates new problems by yet again ignoring the contingent situatedness of human practice in a highly complex and ever-changing world. This 'paradox of technology'[13]—to loosely follow the thoughts of Luhmann (1993)—is amplified in an increasingly economised environment, in which technology is used as means to control economic risks. Thus, investments in new technologies provide solutions to problems by opening up new risks and uncertainties that seem to call for new investments. It is within this oscillating between technological enframing, daily practics and economisation that one must understand the socio-technical reshaping of 'small' vessels into increasingly economised harvest machines.

All in all, the case of the haddock-crisis has vividly illustrated the tensions and dilemmas economic actors face when coping with multiple sources of uncertainty stemming from a highly volatile and more-than-human market environment that is prone to undermine long-term economic investments and decision-making. Today, Bjartur has accepted this new market-driven flexibility as part of his life as an independent small-boat owner:

> It's very complicated … Maybe after five years, then we have a totally different position, you know, then we maybe have no cod and a lot of haddock, because it's always changing and you have to control your company with that in mind that it's always changing and you have to find a way, which is a good way so the company will survive, you know what I mean? (MII)

This quotation not only illustrates the constant struggle for survival of small-boats, in which short-term profit orientation turns out to be a dead-end. It also points out that staying afloat in the industry means constant adaptation to a multiplicity of changing conditions. After all, one might question whether it will ever be possible to come up with

---

[13]For a more comprehensive elaboration on the 'paradox of technology' in the context of market-based resource management see Dobeson (2018: 11–13).

a perfectly stable management regime in the fisheries. This chapter has shown, however, that the solution certainly cannot be found in science and technology itself.

While our journey to the world of fishing has shown us how fishers, quota owners and processors struggle to stay afloat in increasingly econ-omised and technisised resource-management regime by successively adjusting and reshaping small-boats by recombining skills and inte-grating new technologies, it continues to remain a bit of a mystery how small-boats could be transformed into a highly profitable business by relying on rather labour intensive and inefficient capture methods. The following chapters will however show that precisely these ancient ways of catching fish have become the backbone for revaluing the coastal fisheries towards a highly valuable, quality oriented market-niche.

References

Acheson, J. M. (2003). *Capturing the Commons: Devising Institutions to Manage the Main Lobster Industry*. Lebanon, NH: University Press of New England.

Aspers, P. (2011). *Markets*. Cambridge: Polity Press.

Aspers, P. (2018). Forms of Uncertainty Reduction: Decision, Valuation, and Contest. *Theory and Society, 2*(47), 133–149.

Bazilchuk, N. (2010). Mackerel Wars. *Frontiers in Ecology and Environment, 8*(8), 397.

Bear, C., & Eden, S. (2008). Making Space for Fish: The Regional, Network and Fluid Spaces of Fisheries Certification. *Social and Cultural Geography, 9*(5), 487–504.

Beckert, J. (1996). What Is Sociological About Economic Sociology? Uncertainty and the Embeddedness of Economic Action. *Theory and Society, 25,* 803–840.

Beckert, J. (2009). The Social Order of Markets. *Theory and Society, 38,* 245–269.

Braun, B. (2005). Environmental Issues: Writing a More-Than-Human Urban Geography. *Progress in Human Geography, 29*(5), 635–650.

Dequech, D. (2003). Uncertainty and Economic Sociology: A Preliminary Discussion. *American Journal of Economics and Sociology, 62*(3), 509–532.

Dobeson, A. (2018). The Wrong Fish: Manouvering the Boundaries of Market-Based Resource Management. *Journal of Cultural Economy, 11*(2), 110–124.

Fiskistofa. (2015). Total Catches of Species in the Icelandic Quota System. Retrieved from: http://www.fiskistofa.is/english/quotas-and-catches/total-catch-and-quota-status/?skipnr=0&timabil=0708&fyrirspurn=Um-Skip&landhelgi=i.

Holm, P., & Nolde Nielsen, K. (2007). Framing Fish, Making Markets: The Construction of Individual Transferable Quotas. In Y. Millo, M. Callon, & F. Muniesa (Eds.), *Market Devices* (Vol. 55, pp. 173–195). Malden: Blackwell.

Knight, F. H. (1921). *Risk, Uncertainty and Profit*. Boston: Houghton Mifflin.

Luhmann, N. (1993). *Risk: A Sociological Theory*. Berlin and New York: Walter de Gruyter.

Muniesa, F., Millo, Y., & Callon, M. (2007). An Introduction into Market Devices. In M. Callon, Y. Millo, & F. Muniesa (Eds.), *Market Devices* (pp. 1–11). Oxford: Blackwell.

RSF. (2015). Haddock Auction Market Price, 2012–2014. Dataset received from RSF. Dataset received from Reiknistofu fiskmarkaða hf.

Whatmore, S. (2006). Materialist Returns: Practicing Cultural Geography in an for a More-Than-Human World. *Cultural Geographies, 13*(4), 600–609.

7

Fishing for Quality

Yeah, I think it's quality now, isn't it – Icelandic fish!
Skipper

Making it white

At the beginning of December the mountains along the fjord are yet again covered in snow and ice. As on so many days during this season, most fishers were forced to stay ashore due to rough seas. This, however, does not mean that there nothing to do, as many tasks need to be organised around the fishing boat. On a day like this, I meet a Bjartur ashore, where he is helping his employees to attach hooks to the newly purchased fishing lines—a way of saving money on the fee the line producer would otherwise add for attaching hooks to the lines.

Besides fishing and helping his employees, Bjartur also takes care of the family's small processing business. Although fishing can slow down for a while during the winter, cold and dry air provides the perfect conditions for traditional dry fish processing. For this reason, we go out in the winter twilight to look at the company's small processing house, in which mainly haddock and seawolf are hung out to produce the traditional dry fish (*harðfiskur*), which still is a popular snack among young and old in contemporary Iceland. Although the bulk of dry fish for

© The Author(s) 2019, corrected publication 2020
A. Dobeson, *Revaluing Coastal Fisheries*,
https://doi.org/10.1007/978-3-030-05087-0_7

the domestic market is produced by bigger companies accelerating the drying process with special mechanised equipment, this artisanal way of processing can still be observed along many fishing communities all around Iceland, the wooden racks calling to mind that this is an ancient tradition.

Bjartur is convinced that this craft of 'drying it slow' produces a much better and tastier quality than the more industrialised way. As with small-boat fishing, this type of processing requires skill and knowledge of the environment to produce a top quality product, he explains while checking on the development of his produce:

> *AD*: So is it important where to place the dry house?
> *Bjartur*: Yeah, the dry house has to be at a good place where it is windy and cold, not on a place where there's no wind, maybe in the shadow or something, and when you fillet the fish and put it on a dry house it is better to have a cold weather, then the fish get more white, when you put the fish outside when it's hot outside, it cannot be as white as when it's cold weather, so it can be maybe more yellow, and we don't want any – we want to have white fish… (MII)

Apparently, the practice of drying fish not only influences the overall taste and quality of the product, but also its physical appearance. I therefore wonder whether this makes any difference to the value of the product:

> *AD*: Do you get different prices when the fish doesn't look so white?
> *Bjartur*: No, you just, you can sell the fish better when it's white…
> *AD*: Ja…I guess it's a part of your reputation to have a good fish?

> Bjartur Jaja, like if you take Norway and Iceland with the salt fish, Iceland is always with the better fish because we always have a white fish and the process to work with the fish, all the process from it's been fish from the sea and when it is sold to the customer. We have good processing and then we can sell it white. And in Norway, they are not selling as well as we are doing, because they are not having it as white as we do… (ibid.)

Image 7.1 'You cannot fool a Spanish housewife!.' Salt fish processor from South-West Iceland demonstrating his premium product, a filet of 'white' salted cod for the Spanish market (FN: 10). The importance of 'making it white' was stressed by both fishers and processors (Photo by AD)

Accordingly, it is the 'whiteness' of the filet that builds the trademark of Icelandic fisheries (see Image 7.1). Hence, he points to an important feature of the economy that has been overlooked by the neoclassical market model, which assumes the homogeneity of products in the market. In other words, the whiteness of the filet signifies an important quality that represents the identity of the Icelandic fishing industry as producer of top-grade quality products that mark a difference to other products on the market (more on this later). Moreover, the quotation also makes clear that the 'quality' of Icelandic fish also signifies a marker in relation to a collective 'We' identity that marks a difference to the constitutive 'Other', in particular the Norwegian fisheries.

Naïve as I am, as an ethnographer I wonder whether this perceived difference between Norwegian and Icelandic fish could be explained by different fish stocks or environmental influences:

> *AD*: But the whiteness has nothing to do with the fish?
>
> *Fisher*: No, no, no, it's the processing! The fish is the same thing in the water, it's how you control the fish, and line fish is better than trawl fish, and the fish with the [jigging] computer is very good, because you take the fish, it comes up right away and it's very well alive when it comes up and you cut the fish, you get the blood out of it, the blood goes out of the body and you put it in ice water, which is very cold and then you have the fish very white and very cold, that makes a huge difference in the process! And when you come with the fish in the fish company, they can make a good fish if we come with the good fish. They cannot make a good fish if you come with a bad fish, you know what I mean? (ibid.)

For Bjartur, it is clear that the whiteness of the fish in fact has nothing to do with the genetic coding of different fish stocks or the environmental circumstances of the fish's natural habitat. Nor can it be explained simply by the knowledge and skills of a single entrepreneurial individual. Instead, we can say that *making it white* is a skilful practice that is grounded in a distinct 'epistemic culture' (Knorr Cetina, 1999) of production. But does increasing economisation imply a downward trend towards lower-quality products in the name of profit-making?

Instead of suggesting an increasing orientation towards economies of scale, this chapter points out how new markets and technology have reconfigured the discourse and practices in the small-boat fisheries by revaluing quality standards towards the collective construction of a new market and highly valuable market niche for 'line caught' and 'fresh' fish. Hence, instead of destroying the century-old tradition, artisanal and labour intensive capture techniques have become the bedrock of profit-making in new markets that allows small-boat fishers to stay afloat in the new culture of liberal rural capitalism.

The chapter is structured as follows: first, the chapter will clarify the give a brief overview on the economic sociological literature on quality construction in markets. Thereafter, it will be demonstrated how new quality standards are negotiated between producers, markets and the fisheries, in which new ways of handling fish and technology play build the foundations for the 'quality-turn' in the small-boat fisheries.

Qualifying Goods in Markets

Standing in the light of the growing field of a sociology of valuation (Antal, Hutter, & Stark, 2015; Beckert & Aspers, 2011; Lamont, 2012), the role of quality in markets has been raised by a number of scholars dealing with the social organisation of markets (Ahrne, Aspers, & Brunsson, 2015; Beckert & Musselin, 2013b; Callon, Méadel, & Rabeharisoa, 2002). Most commonly, 'quality' is defined in contrast to 'quantity'. Whereas quantity merely involves the amount of something, quality refers to a differentiation according to a standard, distinct property or attribute.[1] But of what nature are these qualities?

From a social constructivist perspective, the 'quality' of fish—as of other goods—is not an intrinsic property of the object itself, but an attribute that is negotiated and defined in a social process:

> 'Quality' is not something that is naturally given, but the outcome of a collective process in which products become seen as possessing certain traits and occupying a specific position in relation to other products in the product space. Hence goods and services become 'qualified'. (Beckert & Musselin, 2013a: 1)

---

[1]According to the New Oxford American Dictionary (version 2.1.3), 'quality' is defined (1) as 'the standard of something as measured against other things of a similar kind; the degree of excellence of something' or (2) as 'a distinctive attribute or characteristic possessed by someone or something'.

Hence, the quality assessment of a fish makes only sense in relation to commonly agreed standards and conventions that allow to differentiate 'high' from 'low' quality products in the market. Prices, however, cannot be used as credible signals of quality, as they can be influenced by market power and status of producers or determined by other factors such as shortages in the supply-chain, which is common in fish markets. Rather, 'quality assessments precede the economic valuation of products and constitute the basis for price formation' (Beckert & Musselin, 2013a: 22). Put differently, objects need to be *qualified* in a set of societal discourses and practices that define quality standards around measurable parameters such as age, size, temperature, texture and density. Once established, 'quality' itself can become a marker for market differentiation that reflects producer's identities and price formation in markets (White, 1981). In this light, we can also understand the rise of 'judgement devices' (Karpik, 2010) such as product labels and certification schemes as ways of qualifying goods as singular objets that differentiate from other goods in markets. In a similar vein, Callon et al. (2002) have underscored the socio-technical aspects of qualification. Hence, 'qualities' are the outcome of a collective process involving not only human values and norms, but also technical artefacts such as digital consumer platforms that allow for reflexive decision-making and market differentiation.[2]

While the economic sociological literature provides important insights into the socio-technical context in which goods are qualified in the product sphere, surprisingly little has been said about how the context of production shapes the qualification of goods over time. For instance, the emergence of mass consumer markets for frozen fish was intimately tied to technological innovations in freezing and storage technologies (Rees, 2013; Thévenot, 1979). Ever since, however, small and medium-sized processors have been unable to compete in these capital intense and economies of scale dominated markets. With traditional niche markets for dried and salted fish in decline and their necks up in debt, however, small and labour intensive producers such as Bjartur's

---

[2]See Chapter 8 for the role of digital technology for quality assessment in fish markets.

family had to re-orient practices of production by requalifying their catcht through the collective construction of a new and highly valuable market niche for 'fresh' and 'line caught fish', as the following will show.

The Quality-Turn

One just needs to ask members of the older generation of Icelandic fishers how fishing was in the often romanticised good old days of the herring rush. They will soon explain that the fishing was literally 'insane' and boats were filled with so much fish that they were often on the brink of capsizing. In the system of open access, so the cliché goes, 'rational' fishers typically try to fish as much as possible in order to get their share of the profits. This may be intensified when fishing in a system with limited days at sea, in which fishers engage in so-called 'Olympic fishing'—that is, highly competitive fishing—until the quota is fished up and the fishery is closed again for the season. Moreover, fishers would not really pay particular attention to the quality of the fish, which were simply lying on deck no matter what temperature it was during the day, as local processors had already set the prices. In other words, quantity dominated quality.

Today, property-rights management regimes such as the Icelandic Individual Transferable Quotas (ITQ) system are often contrasted with systems characterised by 'Olympic' fishing. In this system, so the theory goes, fishers are under less pressure to put to sea because their quota share is individual and set for a fishing season, leaving the fisher to decide when to put to sea.[3] Although fishers—or in this system, the quota-owners—are thus disentangled from their community ties, we have already seen that the process of economisation tends to re-entangle them in a web of money-mediated relations that reconfigure the economic expectations more and more towards profit-making (see Chapter 3). One way of coping with the increasing pressure towards

---

[3]That this is not necessarily the case in a market-based system has already been pointed out in Chapters 4 and 6.

economisation are investments and the rationalisation of operations by means of technological adaptation and risk-taking (see Chapter 6). Another coping strategy for dealing with the increasing pressure towards profit-making is increasing the value of the catch, as Bjartur explains:

> And this is such small place, the names are always shown when they are selling fish on the fish market, and the buyers, if they have a bad fish, they just say 'Fuck you!' [laughter]. So, and all fishermen in Iceland, especially from the small boats, they want to come with the good fish, good big fish, so you can have a good price for fish because if you have ex-quota maybe 100 tonnes, it makes a huge difference to have a good price for these 100 tonnes you have…In the end, if you come with a good fish, maybe you get 3 million more and if you have come with a bad fish… (MII)

Bjartur makes clear that the focus on quality is not a way of maintaining ties with buyers, but also of constantly achieving a higher price at the fish auction by delivering in accordance with the quality standards demanded by the buyers. Hence, fulfilling these standards can be understood as a strategy of coping with the scarcity and limited availability of fishing quotas by means of quality-upgrading: fishers simply try to achieve the highest market price for their catch. But how do the fishers know what quality their buyers demand?

Teaching Quality

Today, the focus on quality has become a mainstream strategy for professional fishers to maximise their profits in the industry. Issues of quality and quality orientation were raised by almost all interviewees when discussing issues such as the quota system or the fish auction as one of the key issues regarding long-term economic stability. In this sense, the slogan 'You have to think quality, not quantity' (FN: 101), indicates not only a changing rhetoric and mind-set, but also a changing practice different from the olden days of quantity orientation that was characterised by low market prices and little care for the catch on board.

That the quality-turn has a direct impact on the practices at sea becomes clear in the following conversation with an experienced processor who has followed the Icelandic fishing industry through recent decades:

AD: Has there been a change since the quota on the small boats, how they treat and handle the fish?

Processor: Yes, that has changed much! When we had a day system, it was an Olympic catching, that was not good to treat the fish, it was very hot when it came in. Now when the fish is landing now, we measure all the heat in the fish, we started last year to measure every - every boat that came in!

AD: You take samples?

Processor: We take samples, we said them what sample would you like us to measure, they chose that and we just okay! And I think everyone is really pleased with that [knocking on the table], as I said earlier: everyone will do his best! (MXIX)

The quotations from the processor also point to a changing practice regarding quality since the implementation of ITQs in the small-boat system. It also points to another aspect of the quality turn: the role of intermediaries.

Fishers do not know what quality is being demanded by simply observing the market. Neither do they all of a sudden admire quality as an intrinsic value of the fishing industry. Instead, quality is constructed in a collective process, in which standards are actively negotiated by a set of different interest groups and passed on to the fishers, as the conversation with the same processor makes clear:

AD: I heard that from other fishermen, the other system was horrible [regarding quality], they say that themselves!

Processor: But there was no one telling them, there was no one forcing them to do that. But now, we put pressure on them: you should take ice with you and you should do it well, and we measure it when it comes in: we do that, no problem! And we have a research for May and June this year [2013], that was pretty good, over 95 per cent of the boats – someone has accidents and things like that, of course that happens, but 95 of the boats are splendid and the fish is around zero [°C], so we like it like that. (MXIX)

The processor points out that he actively engages in communicating quality standards and even puts pressure on the fishers to conduct certain practices such as taking ice on the boats to keep the fish chilled over the summer months. In order to do so, processors and local auction markets either provide ice for free or against a small fee to facilitate the delivery of chilled raw materials.[4]

The processor also points towards other actors who co-construct the discourse on quality: the Icelandic Food and Veterinary Authority (Matvælastofnun or MAST) and Matís, a government owned R&D-oriented agency dealing with food-related issues. These organisations have gained importance in the discourse on quality only in recent years, especially because quality issues have been reported by processors since the implementation of the part-time summer fisheries. According to some producers, many fishers did not cut and store their catch correctly in the beginning—an aspect often used by critics to argue against the system. To improve the system, however, the National Association of Small Boat Owners (NASBO) and processors engaged in cooperation with the aforementioned government agencies to control and educate fishers to improve quality-related issues.

The role of the MAST lies in the measurement and control of landings at local fish markets. For this purpose, local representatives take samples at auction markets on a regular basis over the summer months, a practice I witnessed several times during my time in the field.[5] Furthermore, researchers from MAST monitor fish processing activities and evaluate the collected data in the form of reports (MAST, 2012).

As a further consequence of poor quality landings, Matís was assigned to developing information material and workshops for fishers to improve the quality of their catch (Matís, 2010). In the beginning, the workshops were organised by Matís employees who travelled around the coastal communities, but today they are organised on a voluntarily

---

[4]Like fishers, local auction markets compete with each other and thus also have an interest in maintaining high quality standards as a basis for their reputation among fisher and buyers.

[5]Controlling the fish also means controlling the market. Once during an interview with the manager of a market (MXVIII), a researcher interrupted the interview and raised some issues about the ice, leaving the interviewee rather upset.

basis by local education centres. An important feature of these programmes are justifications and scientific explanations of the importance
of quality, as displayed in the information brochure *Mikilvægi góðarar
meðhöndlunar á fiski* (The importance of good treatment of fish) (Matís,
2010). In order to explain 'good' treatment, 'bad practices' are contrasted with 'good practices', illustrated with images and photographs
of how to use ice and the 'right' cutting of fish—a heuristic method also
used in other publications on quality issues (e.g. MAST, 2012).

Although these education programmes and campaigns were directed
mainly towards part-time fishers, there is no doubt that the involvement
of government institutions has changed the attitude towards an increasing awareness of quality issues among fishers. But what practices result
in the production of 'top-quality Icelandic fish'?

Catching Quality

The physical features of codfish can depend on its location, which influences important parameters such as sea temperature, food chain or
parasitic infestation (also see Chapter 8). It is, however, not enough to
simply catch 'good' fish. As the meat of fish is a very sensitive raw material, the process of qualification therefore begins right at the beginning
of the value chain and is influenced heavily by the socio-technical context it is harvested. For instance, while fishing with dynamite may be
an effective way of harvesting fish, it is not necessarily a good one when
taking the quality of fish into account, as fish are likely to be damaged
by the explosion and shockwave. When fishing with a single hook, the
physical condition of the fish is likely to be very different, as it is usually
landed alive without any bigger physical damage to the meat.

In the Icelandic fishing industry, fishing gear is an important marker
that differentiates raw materials in markets according to different quality standards, as the established distinction between trawl- and *net-caught*
fish and *line-caught* fish makes clear. Accordingly, line-caught fish is
thought to be of much better quality due to the ways this method affects
the physical features of the raw material 'because the fish in the trawler is
squeezed and it's more red in the muscles, it's dead when it comes up in
the trawl, most of it because they are catching so deep while we cut it all

alive' (MVI) as one fisher explains. Hence, small boats usually do not fish as deep as the large industrial trawlers, which can harm the inner organs through changing air pressure when being hauled up; they also use hook-and-line based techniques that shorten the timespan between hook-up and landing. Furthermore, fishing lines are usually retrieved shortly after they have been set. While the fish are hooked, they can—in contrast to being captured in a net—move fairly freely without getting squeezed or otherwise damaged. Fishing with jigging computers, furthermore, shortens the timespan between hook-up and landing to a minimum, as the fish are brought up by the jigging machine almost immediately.[6]

It becomes clear the way in which the fish are captured plays a crucial role with regards to conventions of 'freshness' and 'whiteness' of the fish, as the fish are captured alive before the blood starts to saturate the filet. Unlike a piece of art or a bottle of wine, fish and other raw materials such as vegetables or meat have a tendency to rapidly decrease in quality over time and pass the stage for processing and human consumption with increasing age as organic decomposition through bacteria already begins shortly after the harvest. For this reason, the freshness of the raw material depends strongly on the time period between catch and processing. To cope with the tension between quality and time, practices have been developed to delay the process and extend the timeframe in which quality is preserve. The treatment of raw materials therefore requires a set of practices and devices to construct and maintain the freshness of the fish, as the next section will make clear.

Caring for Quality

While different capture technologies impact the physical qualities of fish in the most profound way, the way it is landed after the catch plays crucial role in maintaining its value. Large-scale trawlers solve this problem by directly processing and often freezing the catch on board, basically producing the end product for the final consumer market at sea.

---

[6]Modern jigging computers even allow the programming of sensitivity.

According to their size, however, smaller boats do not have this option and need to try to land the fish as fresh as possible for the processing plants. Hence, the crew needs to *care* about quality, which implies a set of routines and practices at sea (also see Chapter 4).

> So it [the fish] comes automatically up [the fish], and what you should do is to [makes gesture of cutting the fish] throw it in a box and throw it in, in here [the hold], with ice, so this is quite advanced and a good way of catching fish, and you get excellent quality! (MVI)

Hence, in order to deliver the highest possible quality, the fishing crew must land, cut, cool down and store the fish. What sounds simple, however, requires not only a lot of tiring manual labour, but also knowledge and skill that is directly tied to the quality standards of the market. This process of qualification is explained in what follows:

Landing the Fish

In hook-and-line based fisheries, caring about quality starts with the gaffing of the fish with the landing hook, and the skill of the fishers lies in safely landing the fish with the landing hook without damaging its most valuable parts (see Image 4.3). In order not to damage the valuable filets, the fisher therefore always aims at the head of the fish, which is of least value. If the fisher misses the head and hits the torso or the tail, however, the hook might destroy the inner organs of the fish and cause internal bleeding, which means the processing house will not be able to sell the fish for the highest. In contrast, unhooking the fish is fairly unproblematic for the quality, although it may require some skill when a lot of fish are coming up at the same time, in particular when jigging for cod.

Cutting It Clean

To ensure the emblematic 'whiteness' of the filet, it is important that the fish bleeds out alive to let the organs pump out the blood from the organs and the muscles. To achieve this, a single clean cut with a sharp knife through the throat of the fish is conducted directly after landing.[7] The experienced fisher will therefore sharpen the knife multiple times during the fishing operation to ensure a clean, fast and safe cut.

'Cutting it clean', however, is not an easy task and requires a lot of practice to achieve swift coordination, especially in rough seas. Hence, a misjudged cut not only slows down or even prevents a clean bleeding of the fish, it may also damage the valuable parts. Part-timers and rookies in particular may struggle with this task, to the chagrin of skippers and processors.

Storing and Cooling the Fish

While the whiteness of the fish is more of an aesthetic criterion important for sales, its temperature plays an important role in maintaining freshness by stopping the development of harmful bacteria after death. Furthermore, keeping the fish chilled also slows down rigor mortis, which plays an important role in quality and texture in processing.

While cooling the fish is not so much an issue during the cold season, it becomes more problematic during the summertime, when temperatures rise above the accepted level. During this time, processors are especially keen on controlling whether the catch has been cooled down and stored correctly. Experienced fishers know about this, as their reputation as reliable sellers depends on fulfilling these quality standards:

> Did you cool it down? Did you put a lot of ice over the fish? Because the summertime it's 20 degree, the sea is 12 degree hot, and when you come and land you want to have the fish maybe 1–3 degree hot, 1–3 degree hot the fish, and then you have to cool it very down, you know, you know what I mean? Because it comes from the water with 12 degrees and yeah, that makes a huge difference. (MI)

In order to reduce damage to the fish from heat, it is important to cool it down immediately, which is commonly achieved by storing the fish

---

[7]Gutting and filleting, however, are mostly left to the processors or fish markets, which often provide this service against a fee for the buyers.

Image 7.2 Cooling down the fish (Fresh cod stored in boxes filled with seawater, ice and blood) (Photo by AD)

in a 'combination of ice and sea water to cool down the fish as fast as possible to guarantee best quality of fish (ibid.)'[8] (see Image 7.2).

We have now seen that small-boat fishers have re-oriented their practices from landing quantity to quality fish. Hence, quality upgrading can be understood as yet another coping strategy that allows fishers to respond to the increasing pressure of economisation by maximising their

---

[8]To improve the quality of fish from small boats, a 3X Technology, a local company from Ísafjörður has developed special machinery for small boats that is supposed to optimise the cooling and bleeding process: after the fish is manually landed and cut, the fish is forwarded through a washing tank where it stays about 10–15 minutes. In contrast to the conventional storage of fish, which is simply lying in a mix of ice, water and blood, the fish is constantly

profits in order to stay afloat. But why would processors pay premium prices for line-caught fish when cheaper alternatives are abundant?

Producing Quality

For over a century, the trademark activity of Icelandic fisheries has been the production of salt fish for southern European markets, in which salted cod used to be in high demand as a basic ingredient of the traditional cuisine. Big line-caught cod used to be the most valuable landing for producing the typically snowy white delicacy, for which especially Spanish, Portuguese and Italian consumers were willing to pay high prices. Today, Icelandic salt fish still holds the reputation of a luxury product that is commonly known and marketed in distinction from other products, such as 'baccalao de islandia' in Spain, which is still one of the main export markets for salted cod. With the enduring economic crisis in Southern Europe and changing consumption patterns, however, the golden era of the salt fish industry seems to have passed, as the demand for cheaper alternatives is putting pressure on prices.[9] However, the salt fish crisis does not mark the end of small boats, who seem to benefit from the devaluation of the Icelandic króna in the aftermath of the financial crisis, as a fisher explains:

> We are selling fish to France and Germany, I think that's a better market. The crisis makes a huge difference, and we have a crisis in Iceland and Iceland króna is getting low, so we get more crowns because we sell the fish to Europe in euro ... (MI)

---

in motion during the bleeding process: similar to a washing machine, a spiral ensures that the fish is cleaned with seawater that is constantly running through the system. According to the manager of the company, the system also delays rigor mortis and therefore the production of bacteria, which is of major importance in ensuring the highest quality when the fish is landed. In 2014, one bigger long-liner from the village of Bolungarvík had already installed the system, while other orders were pending. At the time of writing, the system seems to be attractive only to bigger longliners of around +15 tonnes in the small-boat system (FN: 92).

[9]Based on interviews with three salt fish producers conducted during a pilot-study in Southwest Iceland in 2012.

This statement makes it clear that the small-boat fishers benefit from the devaluation of the króna in the post-crisis economy, which has opened up ways of reorienting the exports to the booming economies of northern Europe. Salt fish, however, is widely unknown to consumers in countries such as the United Kingdom, France or Germany. For this reason, a new market niche had to be invented to compensate for the declining salt fish business.

The Booming Fresh Fish Market

For this reorientation of the Icelandic small-boat economy, the emergence and boom of the fresh fish market has been of particular benefit to small boats caterin small- to medium-sized processors who are increasingly shifting from salt-fish processing to a high quality niche for fresh chilled fish (see Appendix, Fig. A.6).

In cooperation with the airline Icelandair, which has been steadily increasing the air transport of chilled fish to Europe and the United States on a daily basis (see Appendix, Fig. A.7), processors can ensure their contractors that the fish can be offered to their customers for high prices in as little as 48 hours from its landing. Due to this increasing demand, fresh fish has become the most valuable product segment that more or less equals frozen products despite significantly lower quantities (see Appendix, Figs. A.8–A.9).

Today, processing for the fresh fish market has become a highly profitable substitute for the declining salt fish business and allows smaller- to medium-sized producers more flexible production than the highly cost-intensive and economies of scale-oriented freezing market, as a processor working exclusively with small boats explains:

AD: Why did this company enter the fresh fish and not the freezing market?

Processor: Okay, [with] the frozen fish normally the price is lower; normally it's like that, you know. But the fresh fish, you have to have a very stable quantity of fish coming in and because you have to supply the costumer, but in the frozen you have hundreds of tonnes in the freezer and just ship it out. Also in the salt fish, it takes 28 or 30 days

> to produce it, then comes some days, you know, two or three weeks in transport and it comes another 30 days in payment, so it's three months almost – when you [knocks on the table] catch the fish until you get the payment! But in the fresh fish, you know, transport fish today and you know, the cash flow is much quicker! (XIX)

Thus, the fresh fish segment allows smaller- and medium-sized companies that do not have the means to buy, process and store large quantities of raw materials to operate on a more flexible, future-oriented basis that make it easier for them to cope with the volatility of supply and prices. In the same way, another processor, whose company shifted from salt fish to fresh fish processing, also highlights the future-orientation as a positive aspect of the fresh fish market for his company: 'We can produce it every day' (XII)—making the company more adaptive to a changing market environment and avoiding costs for refining and storing the produce. Thus, many smaller- to medium-sized companies engage in flexible just-in-time production to cope with the fluctuations that characterise the fishing industry.

The same processor also points out why the company is focusing exclusively on the fresh fish market and not—like some of the large-scale processors that have begun entering the fresh fish market—engaging in both frozen and fresh fish processing by using the most valuable part, the loins, for the fresh fish segment and the rest for the freezing segment: 'We don't have the freezing machines to do it. ... This costs a lot of money to do that' (ibid.). The processor points out, however, that this is not necessary or worthwhile for the company because 'the price [on the freezing market] is that lower, and you know, the demand for our fish is so high, so we don't have any [incentive]—it's that much that we [can barely cover the demand]' (ibid.).

Despite booming exports, however, getting involved in the fresh fish market is not a suitable strategy for everyone in the industry, as he continues to explain:

> For example, you know, if guys in the northern side of Iceland, they gotta transport one, maybe two tonnes every day before two o'clock to the airport – that's quite different – difficult for them! So it's much easier for

them just to put the fish [into the freezing house] because of the weather! Sometimes [there] is, you know, much snow and maybe [the trucks] cannot come for two or three days and the costumer is very unhappy and, you know! It's unstable! (XII)

Thus, companies concentrating on the fresh fish segment are usually located in the capital region or the Southwest to ensure regular access to airport facilities. In contrast, the Westfjords in particular tends to be precarious and unstable with regard to weather and poor road conditions, which make it risky to invest in such a business. This does not mean that fishers based in remote rural regions are not affected by the demand for high quality fish, however, as they serve this market segment through the national fish auction (see Chapters 3 and 8). At the time of writing, however, no processing company from the Westfjords region was specialising exclusively in production for the fresh fish market. While some companies still serve the traditional markets for dried and salted fish, a new market niche has emerged that serves the increasing international demand and willingness to pay much higher prices for 'line-caught fish'.

Authentic Protein

Marketing research has shown that capture technology accounts for product differentiation in final consumer markets (Grundvåg, Larsen, & Young, 2013). Accordingly, the attribute 'line-caught' accounts for a price premium in UK supermarkets of 18% for cod and 10% for haddock, compared with other fishing gear. Thus, consumers are willing to pay more for fish caught by 'sustainable' and 'environmentally friendly' capture technologies than for fish caught with conventional industrial gear, such as trawls and gillnets. Hence, fishing gear as become an important marker of market differentiation, which equips the fish with a special singular quality that is reflected in the premium price.

Increasing consumer demand and willingness to pay for 'line-caught fish', however, is not only the result of changing consumer values in rich countries, but also the result of a restructuring of processing companies and a marketing campaign that constructs and places the quality of line-caught fish as an environmentally friendly and sustainable alternative to other conventional 'mass products' in the product sphere. For instance, brands such as *Icelandic* market line-caught fish from Iceland as top-quality premium products[10] (Icelandic, 2015), and *Demeter*, a German certifier for 'biodynamic products' exclusively markets Icelandic line-caught fish from coastal vessels with the slogan '*natürlich. sozial. nachhaltig*' ('natural. social. sustainable'), referring to what they believe are the socio-economic and environmental benefits of coastal fisheries as part of their marketing strategy (Demeter, 2015). These certifications and marketing campaigns differ from conventional eco-labels such as the Marine Stewardship Council (Gulbrandsen, 2009), as they highlight the uniqueness, quality and ecological and social sustainability of line-caught fish as distinct from conventional mass products.

The boom in line-caught fish, however, is not merely the result of strategic marketing campaigns, but must be seen in light of the broader transformation of the fishing industry and its consequences for smaller coastal communities. Thus, especially in the structurally weak rural areas, such as the Westfjords, fishers, quota-owners and producers needed to reorient their market strategy and positioning to keep their companies afloat. Hence, investments in fishing quotas and new technologies, such as processing equipment and fishing vessels, have opened up new ways of processing and marketing based on local knowledge, as the CEO of a small boat-based processing plant explains:

> And the most change was, I believe, that this company is not owned anymore by some sales company in Reykjavík, but by individuals here, and we changed the strategy that we were on market prices and that means the guys on the boats always get the highest price for the fish, and of course we get the highest price here [for the processed fish], because the thinking, the thinking of how things here were more or less: 'It doesn't matter, this is just the fish!' We changed the thinking about the fish, it

[10]'In today's demanding market place, we excel in the quality of our chilled, line-caught fish' (Icelandic, 2015).

is food! We are not just a processing plant, we are a processing plant for high quality fish, it started here! (XIX)

Hence, the company has changed its strategy from 'conventional processing' to serving a 'high quality'-oriented market niche that has enabled an upgrade in the value chain for avoiding competition with standard-oriented mass markets (Aspers, 2010). With the quality-oriented processing, the company could now serve a different market niche for exclusive fish products that can be sold at high prices in wealthy industrial nations such as the United Kingdom, Germany and France. In addition to this quality-upgrading, the company took part in a marketing campaign that not only highlights the 'quality' of the raw material as such, but also other qualities referring to the ethical dimension of small-scale industries. Accordingly, fisheries using artisanal and hook & line-based capture techniques not only provide a better raw material for processing and consumption, but also create jobs for the coastal communities and allow the maintenance of a sustainable fishery in distinction to the large-scale industrial fleet. That these discourse are now deeply inscribed in the corporate identities becomes clear in this statement by another processor working exclusively with small boats:

> Because you don't use as much oil on this one [coastal vessels] is because you don't go as far, first of all. And you don't use this huge power to drag the trawl, you know, on the ground. And you don't spoil the bottom like the trawl is doing and, it's been calculated that we are using between 20–25 króna per kilo cost for oil, while they do like 40–50 króna per kilo on the trawler. And of course, we don't pollute as much then with CO_2, so it's environmentally friendly in many ways. (VI)

Today, many smaller- to medium-sized processors benefit from the increasing boom for high quality produce from small and artisanal companies, and many consider the new focus on quality as a win–win situation for both fisheries and consumers: 'You know what you buy, you are buying good quality and I have a good price for it because this is just *real* food' (VIII). This statement makes clear that line-caught fish is not only one product among others in the product sphere, but stylised as a more authentic source of protein that stands in opposition to the unsustainable and artificial world of mass production. It is for this reason that line-caught fish from coastal fisheries has become a highly valuable singularity in the product sphere.

The Valorisation of Line-Caught Fish

The changing processing and marketing strategies of smaller to medium-sized producers have not only created new market segments for fresh line-caught fish, but have also boosted the domestic demand for fish from small-boats. Although some processors have their own small fleets with one or more longline vessels, the demand is usually much higher, so that raw materials need to be bought at the local auction. In particular, companies serving the fresh fish market are highly dependent on the supply of line-caught fish due to the higher quality of the raw material, as a processor explains:

> *Processor*: Usually we just buy it from the line-caught [segment], that's the only [kind] we [use].
> *AD*: But do you do marketing with this, having this kind of corporate identity? Or is there a demand for that from your costumers?
> *Processor*: Usually there's a demand, because the fish is much whiter, the – what you call it? – the fish itself is not, but the filet is much whiter than the net fish because the blood has been over the filet…
> *AD*: So is it better quality?
> *Processor*: Much better quality! That's why we always have line-caught fish! (MXII)

The attitude of the processing company towards line-caught fish is not merely a niche preference by a single company, but reflects a general high demand for fish, especially for producers serving the fresh- and line-caught markets. This is also reflected in the auction prices, where hook and line-caught fish usually achieve a higher price than other, more efficient capture technologies such as gill nets and trawls 'because it's more fresh when it's line fish' (MXVI). Accordingly, buyers can see at auction what fishing gear was used before they make an offer, usually leading to higher prices for line-caught fish compared with trawl fish, according to an auctioneer (MXVII).

All in all, the booming demands for line-caught fish has led to a steady valorisation of catch in the hook-and-line quota class since the year 2000 (see Appendix, Fig. A.10). Thus, to ensure the delivery of a top quality product and maintain stable ties with buyers and high prices, especially companies involved in the fresh fish market depend exclusively on the supply of line-caught fish for processing. Likewise, companies marketing their products as 'line-caught' are obliged to provide a product caught with hook and line-based fishing gear.

Especially during the wintertime when weather tends to turn rougher, however, the stable supply of raw materials can be problematic, leading usually to the highest prices at auction (see Appendix, Fig. A.11). But even during the summer months, when a lot of part-time fishers are flooding the market with fish, supply is usually short and processors are willing to pay high prices to ensure a stable supply to their costumers abroad.

Conclusion: A New Culture of Quality

While revisiting Bjartur's homeport in the summer of 2018, I was surprised to witness that the last signs of the century old tradition of artisanal dry fish processing I was proudly introduced to a few years ago had vanished from the community. Instead of preparing and putting up a small share of the vessel's catch in the wooden drying house by the fjord, the company's entire catch is now sold directly to much more profitable fresh fish market. While it seems as old forms of knowledge and skill seem to be slowly but surely surpassed by new ways of handling and processing quality, small-boats artisanal and labour intensive capture methods have become the bedrock of profit-making in new markets.

In order to understand the dynamics behind this transformation, this chapter has demonstrated that the 'quality' of goods can neither be reduced to an inherent feature of the object itself, nor something artificial created by marketing experts. Rather, the qualities of goods are qualified within a historically contingent web of socio-technical practices, in which ways of 'doing quality' are negotiated, grounded and reproduced. Hence, in today's fisheries, new ways of handling fish have become widely habitualised and engraved in a unique 'epistemic culture' (Knorr Cetina, 1999) that is structured around specific socio-technical devices and practices that qualify 'line caught fish' as authentic source of protein in premium markets. The re-qualification of line-caught fish to a luxury good not only requires a changing mind-set in processing, transport and marketing, but a 'quality turn' that transforms the practices and cultural identity of the small-boat sector itself.

Although fishers have a general economic incentive to deliver high quality goods, the auction market seems to have put buyers in a rather weak position. Not only are many boats unknown to buyers who can bid from around the island. Storms or stock predictions can shorten the supply, leading to sky-high prices on the auction market. But how can processors be certain that the premium prices paid for the raw materials at the auction market are up to their standards? To answer this question, the next chapter will turn to the fundamental problem of quality uncertainty in markets.

References

Ahrne, G., Aspers, P., & Brunsson, N. (2015). The Organization of Markets. *Organization Studies, 36*(1), 7–27.

Antal, A. B., Hutter, M., & Stark, D. (Eds.). (2015). *Moments of Valuation.* Oxford: Oxford University Press.

Aspers, P. (2010). Using Design for Upgrading in the Fashion Industry. *Journal of Economic Geography, 10*(2), 189–207.

Beckert, J., & Aspers, P. (Eds.). (2011). *The Worth of Goods: Valuation & Pricing in the Economy.* Oxford: Oxford University Press.

Beckert, J., & Musselin, C. (2013a). Introduction. In J. Beckert & C. Musselin (Eds.), *Constructing Quality: The Classification of Goods in Markets* (pp. 1–23). Oxford: Oxford University Press.

Beckert, J., & Musselin, C. (Eds.). (2013b). *Constructing Quality: The Classification of Goods in Markets.* Oxford: Oxford University Press.

Callon, M., Méadel, C., & Rabeharisoa, V. (2002). The Economy of Qualities. *Economy and Society, 31*(2), 194–217.

Demeter. (2015). *Fisch aus nachhaltiger Küstenfischerei*. Retrieved from http:// www.felderzeugnisse.de/TK-Fisch.107.0.html.

Grundvåg, G. S., Larsen, T. A., & Young, J. A. (2013). The Value of Line-Caught and Other Attributes: An Exploration of Premiums for Chilled Fish in UK Supermarkets. *Marine Policy, 38,* 41–44.

Gulbrandsen, L. H. (2009). The Emergence and Effectiveness of the Marine Stewardship Council. *Marine Policy, 33,* 654–660.

Icelandic. (2015). Products & Markets. Fresh Fish. Retrieved from http:// www.icelandic.is/icelandic/products-markets/fresh-fish/.

Karpik, L. (2010). *Valuing the Unique: The Economics of Singularities*. Princeton: Princeton University Press.

Knorr Cetina, K. (1999). *Epistemic Cultures: How the Sciences Make Knowledge*. Cambridge, MA: Harvard University Press.

Lamont, M. (2012). Toward a Comparative Sociology of Valuation and Evaluation. *Annual Review of Sociology, 38,* 201–221.

MAST. (2012). *Hitastigsmælingar á lönduðum afla*. Retrieved from Selfoss.

Matís. (2010). *Mikilvægi góðrar meðhöndlunar á fiski*. Retrieved from Reykjavík: http://www.matis.is/media/matis/utgafa/Mikilvaegi-godrar-medhondlunar-a-fiski.pdf.

Rees, J. (2013). *Refrigeration Nation: A History of Ice, Appliances, and Enterprise in America*. Baltimore: John Hopkins University Press.

Thévenot, R. (1979). *A History of Refrigeration Throughout the World*. Paris: International Institute of Refrigeration.

White, H. C. (1981). Where Do Markets Come From? *The American Journal of Sociology, 87*(3), 517–547.

8

The Fishery Panopticon

> *There are no more secrets!*
> Skipper

Back in the good old days before the quota system, coastal fishers were embedded in local networks of production. Typically, small boats put to sea and landed the catch at their homeports, where the fish was weighed and forward to local processors who evaluated quality and dictated prices. Access to the fishing grounds was free and fishers did not care as much about the quality of their landings, leading to rather low prices in the overall value chain. These rural networks of production, which describe the ideal typical 'producer market' (White, 2002) were based on a clear temporal—and spatial positioning of resources and information that were trickling 'downstream' to processing plants, international distributers and final consumer markets. After all, the fishers built the weakest link in the value chain, being left to the fluctuations of international trade and the monopoly of their buyers.

Today, as we have learned by now, the situation has changed dramatically for independent coastal fishers such as Bjartur who can now decide freely when to put to sea and expect a high price for his catch as multiple bidders from all over Iceland compete over scarce resources

© The Author(s) 2019
A. Dobeson, *Revaluing Coastal Fisheries*,
https://doi.org/10.1007/978-3-030-05087-0_8

on the auction market (see Chapter 3). Within this new regime of market-based coordination, Bjartur even has the chance of striking an exceptional deal, for instance when bad weather on the east or south-west coast cut short supply on the auction market. Hence, it almost seems as the coastal fishers have gained power over the processing sector, as they build the backbone to supply and quality of raw materials. Being subject to the fluctuations of the auction market, however, also implies that fisher may loose out if the market weather is good and many boats are flooding the market with raw materials. Furthermore, with the development of a quality-oriented market niche (Chapter 7), processors have become very selective in regard to the resources they buy. As Bjartur makes clear:

> And when the boats come and land, then the buyers know where they were fishing, and they buy from the boats that was more outside with the bigger fish and they also know who was coming with fish that was cooled down, that was one part where they were fishing and one part was how do you, how do you say it, how do you take care of the fish on the boat. (MII)

It seems as processors are very keen on ensuring to acquire only the highest grade for maintaining the high quality-standard required in the booming fresh fish market. But how can buyers gain access to all this information before they strike a deal?

This chapter argues that we need to reconsider our conception of rural production networks with regards to the emergence of digital communications technologies such as Automatic Identification Systems (AIS) and online tracking websites. We will take a closer look at the relation between markets, technology and fishing in order to understand how and to what extent the temporal and spatial distribution of information and resources has changed economic coordination in rural production networks. We will see that a new, more dynamic 'scopic' valuation regime has emerged based on the remote real-time observation of fishing activities, consequently changing successive downstream-orientation of traditional production networks to increasing temporal and spatial synchrony. As a consequence of this constant surveillance and control of fishing activities, fishers have become subject of an all-encompassing fishery panopticon that pushes them to discipline their practices with regards to the expectations of 'the market'.

The Problem of Quality Uncertainty in Fish Markets

Today a joint Icelandic fish auction brings together buyers and sellers from all over the country by means of a digital bidding platform. Like all markets, however, electronic markets bear the fundamental problem of valuation: due to 'informational asymmetry', buyers often do not know if the price is an adequate representation of a good's quality. The classical example for this dilemma in the economics literature is the market for second hand cars in which quality and prices will deteriorate if not regulated by institutions that force sellers to provide adequate information (Akerlof, 1970). The problem of quality uncertainty is in particular apparent in markets for highly perishable products such as fresh fish, in which handling and time are key to maintaining the value of the raw material. In traditional auction markets this problem is solved by examining the quality of raw materials 'on the spot'.[1] But how do buyers in electronic markets who cannot directly assess quality on the spot make sure that the raw material is up to their quality standards?

Graham (1999: 206–207) sees an advantage in the network structure of Icelandic society for implementing digital fish auctions with remote bidding schemes. Thus, the emergence of remote fish auctions cannot be explained by technological developments alone; the dense network of Iceland's homogenous society and industry are also responsible for the emergence of remote bidding systems based on description and trust. By contrast, the fishing industry in the United Kingdom remains too diverse, complex and specialised, consisting of different types of supply chains, products and buyers, so that the emergence of trust-based

---

[1]A brilliant analysis of the institutions, rituals, signs and practices deployed for overcoming the fundamental problem of quality uncertainty in spot-markets for fresh fish can be found in Bestor's (2004) seminal ethnography of the Tsukiji whole sale market in Tokyo.

relations as a prerequisite for remote auctions remains quite difficult. In the following, however, we will see that this network-based explanation falls short of understanding how buyers and sellers can account for trust in spatially dispersed markets such as the Icelandic fish auction.

Although network ties may have been important for the implementation of the auction system, the following will reveal that this development was much facilitated by the rise of new communication technologies that have opened up new pathways for practices beyond the locally bound and interaction-based networks of the community. Accordingly, the rise of digital information technologies has opened up new pathways for practices by enabling market actors to identify the vessel type, name, position and course of any given vessel based on real-time information. As a consequence, 'scopic' forms of valuation allow market actors to remotely observe fishing activities and estimate the quality and potential availability of a vessel's catch depending on its positioning at sea. This valuation regime does not only change the form of interaction from face-to-face to face-to-screen based coordination and replace the locally bound and successive temporal logic of market networks with a more globalised, flow-like exchange of information (Knorr Cetina & Bruegger, 2002); it also alters the temporal and spatial structure of the economy by increasingly tying fishers into an all-encompassing *fishery panopticon* of remote surveillance, self-discipline and control. In the following we will see how the co-construction of this new valuation regime disciplines fishers to control their vessel's virtual identities and practices to meet the expectations of potential buyers.

Scopic Fisheries

The socio-technical backbone to the construction of an all-encompassing quality regime builds a technology that was neither intentionally designed nor implemented as market device: the so-called Automatic Identification System (AIS). Originally, AIS technology was designed to improve safety and avoid collisions at sea by providing information about vessel positioning, course, type, speed and identity to other vessels and domestic coordinators of marine traffic.

The difference between AIS and regular radar or GPS navigation is that AIS does not merely receive information on the geographical positioning of a vessel and its surrounding traffic, but provides information on its own positioning to a communication network, which allows remote tracking and communication between other AIS-equipped vessels and third parties ashore. In combination with 'scopic media' (Knorr Cetina, 2003: 8), that is, instruments 'for seeing and observing' such as chart plotters and computer screens equipped with special navigation software, the transfer of data thus enables not only the display of other AIS-equipped vessels, but also the storage and recall of navigation data (Image 8.1).

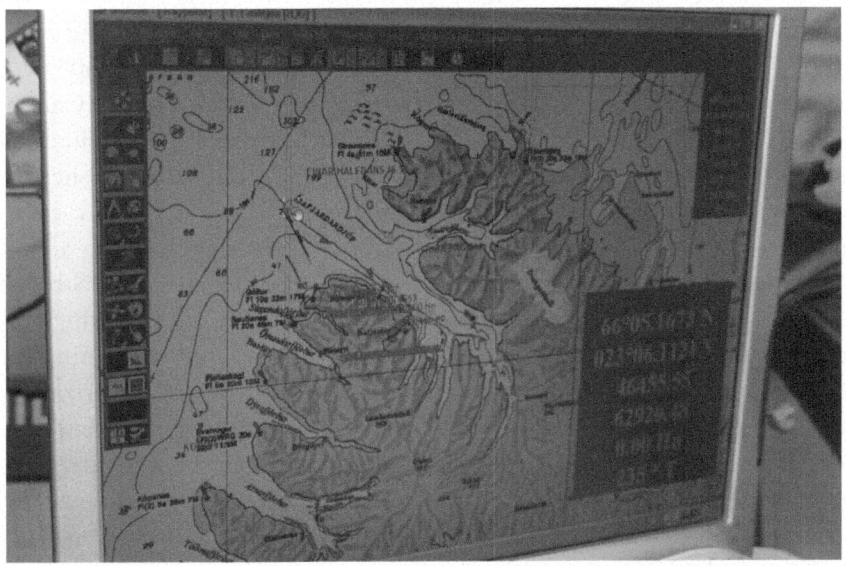

Image 8.1 AIS-data display. Personal computer screen equipped with AIS-software (MaxSea) on board a small fishing vessel in the Westfjords, Iceland. Besides positional information and sea structure, the screen displays information on the current position of other vessels and historical data on the vessel's former fishing spots (Photo by AD)

According to the Directorate of Fisheries (Fiskistofa, 2013), AIS transponders have been obligatory on Icelandic vessels since 2008, when the AIS network replaced an older monitoring system based on automatic radio signals that was implemented in 1998. Today, AIS network seems to cover most parts of the coastal waters, but especially rural areas, such as the Westfjords, where, depending on weather conditions, in numerous fjords surrounded by mountains, reception can be hindered. If a vessel drops out of the system, the coastguard will call up the vessel by radio to check on its current state.

The Facebook of the Sea

While research and reports have highlighted the importance of AIS technology for safety-related matters and monitoring (e.g. Dziewicki, 2007; European Commission, 2012; Norris, 2007; Tetreault, 2005), no attention has yet been paid to its meaning as 'market device' (Callon, Millo, & Muniesa, 2007). Today, the development of public vessel-tracking websites based on AIS technology not only allows fishers to observe each other, but enables observers spatially separated observers to track vessels based on real-time information of their positioning at sea. Put differently, in the Icelandic fishing industry, the use of AIS-based technology has become not only a technical standard on all fishing vessels, but also an important means of observing and communicating for other parties related to the fishing industry since the rise of privately-run vessel-tracking websites. In particular, the website marinetraffic.com, which is run by Maltenoz Limited, a Cyprus-based software company, has become very popular and the dominant vessel-tracking site in Iceland.[2] In contrast to the original purpose of AIS, however, marinetraffic.com does not see itself explicitly as an information portal for accident prevention (Marinetraffic.com, 2013).

---

[2]During the period of fieldwork in summer 2014, however, several fishers started to use the newer vessel-tracking website vesselfinder.com, which, in their opinion, provides more accurate data and 'more information' (FN: 97). When returning to the field in 2017 and 2018, however, the majority of people involved in the fishing industry still seemed to rely on the much improved marinetraffic.com website.

The success of marinetraffic.com may be explained by the fact that the basic functions and recalling of information are free of charge and do not require—as with other vessel-tracking sites—user registration to get started. Users basically type the address into their browsers and can immediately see what is going on at sea. The website has integrated the Google Maps API, on which vessels are displayed according to the information provided by their AIS transponder. Handling the website is very intuitive and allows for quick orientation and immediate retrieval of information (see Image 8.2).

The basic functions are filtering vessel types and searching for specific areas, ports, areas and vessels. When clicking on a vessel, a sub-menu with basic information on the vessel opens up, and further information

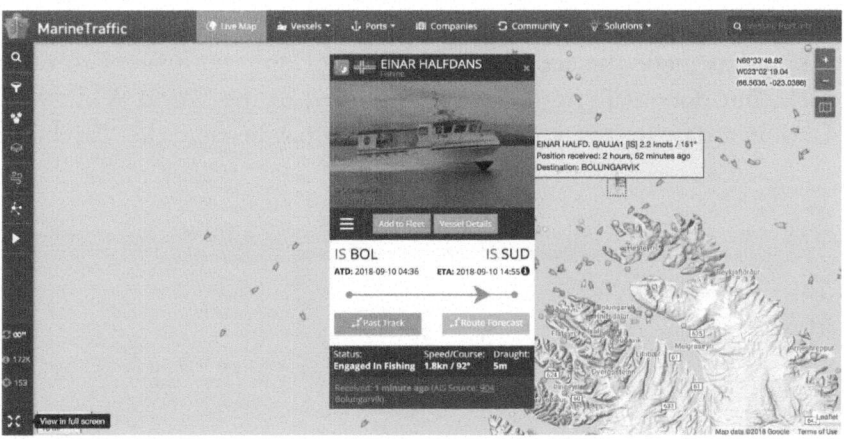

Image 8.2 Synthetic space I. The toolbar to the left allows filtering of the type of vessel to be displayed. The search function to the upper left enables the user to find a specific area, port or vessel. By clicking on a triangle, the user can reveal the identity of each vessel, in this case the Icelandic long-line fishing vessel Einar Halfdans from Bolungarvík. The menu provides information on the flag, ship type, status, speed/course, length and breadth, vessel class and time of the latest synchronisation with the AIS-transponder. If the vessel owner has uploaded one, a picture of the vessel—as in this case—will be displayed with the head menu. Most profiles are equipped with profile pictures, indicating that vessel owners care about their online presence (*Source* marinetraffic.com, 2018)

can be retrieved, for example, on the vessel's speed and current and past course (see Image 8.3).

Users also have the opportunity to create a free account that provides extended functions, such as long-time observations of selected ports and vessels and even SMS alarming, for example, when a vessel of interest leaves its port or enters and leaves a specific area. Moreover, marinetraffic.com has expanded its services by launching a smartphone application that enables the mobile observation of vessels.

In Iceland, fishers make a point of registering their vessel at marinetraffic.com and often also upload pictures of their vessels. During the period of fieldwork, literally everyone within the fishing communities—fishers, buyers of raw fish, processors, harbour managers, fish line baiters, family members, people in the tourism sector (and indeed one ethnographer)—were engaging in online vessel tracking and synchronising their land-based activities accordingly. Hence, the rise of online vessel tracking has made the activities of fishers not only transparent to other vessels, but potentially to everyone connected to the World Wide Web, 24 hours a day. In a way, marinetraffic.com has become the 'Facebook'

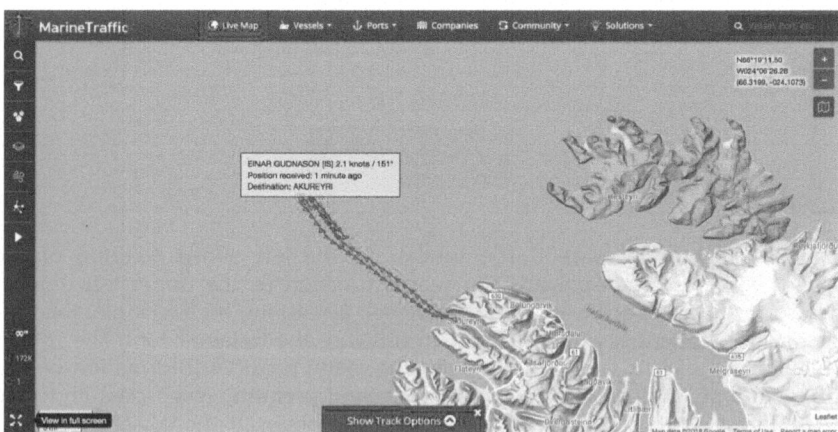

Image 8.3 Synthetic space II. Marinetraffic.com allows the display of the past course of each vessel, here of the fishing vessel Einar Gudnason. Hence, even when the ship is on its way back from the fishing grounds, any user can easily trace back where the vessel did leave its lines (*Source* marinetraffic.com, 2018)

of maritime affairs.[3] In contrast to it social media counterpart, however, the website allows for no direct interaction in the form of messaging or postings between users. Rather, it remains an interaction-free synthetic space in which fishers watch one another, consequently redefining the spatial and temporal nature of competition at sea.

Watching One Another

Although the Icelandic fisheries are no longer based on open-access competition at sea and regulated through a strict quota system, observing other fishing vessels remains crucial for keeping track of their competitors and gaining access to information on potentially productive fishing grounds. In this light, the emergence of AIS has led to the development of new ways of seeing and observing in the fisheries. Today, AIS-based chart plotters and computer screens are standard equipment, even on the older and smaller vessels of the fleet. This is seen as indispensable for success by many fishers, as the following episode makes clear:

> *AD*: You cannot run a profitable boat without a computer?
> *Skipper*: No…
> *AD*: Aha… So what do you need a computer for?
> *Skipper*: You can get much better maps and, and things and do things what you cannot do, you know, then you can have the AIS system, then you can get that up, on the screen and then you can get all the boats up and name of every boat. (MIX)

Hence, in contrast to the old radar system—which is still obligatory on every boat—AIS allows fishers not only to see what surrounds them in

---

[3]Nonetheless, there were some difficulties with implementing full coverage at the beginning. Especially small fishing communities in rural areas had problems due to bad reception, as they needed 'some equipment on the mountains here to catch the signal from the boats' (MXIX). According to a manager of a processing plant who is using AIS-based online services for his business, the reception 'is very good now, it was not good [before]; it was not receiving every signal from every boat' (ibid.), but 'we started to use it every day three years ago (2010) or so'.

a range of up to 12 miles, but to track the exact position, name and course of any vessel fishing in Icelandic waters. Furthermore, special AIS-based software allows skippers to keep a logbook of their former positions and fishing spots, which can be retrieved at any time. Consequently, digital vessel-tracking technology has not only changed navigation at sea, but has also revolutionised the ways in which information is stored and distributed among skippers. In contrast to analogue days, when fishers could hide information on a productive fishing ground or be part of a cooperative between fishers that shared information by word of mouth or via wireless, AIS-based systems have created almost full transparency between vessels, as the same skipper makes clear (ibid.): 'You have no secrets—not for long, maybe just for one day (…). Now everybody knows my secret spots [laughter]' (MIX). Another skipper adds: 'You can go on the computer and you can see where all the boats are, no secrets anymore. Just if you find someone, you have to wake up early to go first, find, go to the spot…' (MI).

Hence, fishers not only make use of AIS-based information sources while at sea, but also rely on AIS-based websites, such as marinetrafic. com, to plan and coordinate their future fishing operations. Especially in larger harbours with bigger fleets, competition for popular or temporarily very productive fishing spots can be very high, and fishers of course try to be as efficient as possible with their fishing operations. For this reason, skippers often get up in the middle of the night to get access to as much information as possible to make decisions about whether or not to put to sea on a particular day.

Furthermore, vessels displayed on scopic media not only provide information on vessel positioning to skippers, but also mirror identities of fishing vessels, as an example from the field makes clear: while explaining what information he retrieves from the internet to evaluate the sea state, a skipper pointed out that he observes other vessels on marinetraffic.com for this purpose. While clicking on a particular vessel on the map to retrieve its name, the skipper reacted with 'Oh, I don't trust this boat' (FN: 667), as it apparently was from a fishing village where the skippers are notorious for taking high risks and being pushed by their owners to go out in rough seas.

The examples make clear that skippers make use of scopic media displaying AIS-based data sources. This, of course, does not mean that skippers do not make use of their network ties anymore to obtain access to more precise information on conditions at sea. Nevertheless, digital scopes seem increasingly to pre-structure information that skippers use in order to frame which ties are worth activating and which not, for example, by identifying vessels that can be relied on to obtain further information after being spotted on the computer screen.

We have shown how AIS-based data sources have changed the nature of competition at sea by linking and synchronising the activities of fishers in synthetic spaces of observation that make nothing but impossible to keep secrets from one another. But what could others gain from tracking vessels at sea?

The Scopic Valuation of Quality

Although market prices are the most important information on which fishers, processors and buyers base their decisions, they have the problem that they do not include any information about the future supply of raw materials. For this reason, both fishers and bidders are keen on gaining as much information as possible on potential supply before the market opens. In the fisheries, however, synthetic spaces of observation now enable actors from different economic domains to track vessels *before* the market opens and new prices are formed. Hence, scopic media not only allow fishers to keep track of other vessels' activities, but allow for 'scopic valuations'[4] of quality based on real-time information of vessel positioning at sea, transforming the temporal and spatial structure of rural production networks towards increasing synchrony.

---

[4]For a more detailed theoretical account of this new form of 'scopic valuation', see Dobeson (2016).

Tracking Quality

While processors cannot directly influence the price or quantity of the raw materials offered on a given day, they can try to control the quality of the raw materials they are planning to acquire instead. In digital markets, grading addresses the problem of quality assessment by allowing buyers in the auction to obtain information on different parameters such age, size and capture technique, which can affect the quality of the raw material. Based on this information, buyers learn by experience which vessels they can trust:

> There is a lot of information you get on the auction: you have the boat's name, for most of the market you have the heat, you have the size of the fish and et cetera et cetera, and with that experience you learn about the boat, you know: it's okay to buy from this one, it is not okay to buy from this one – you will see! (MXIX)

Put differently, links between buyers and sellers are not anonymous, but involve a history of past transactions that bestow a market identity on a vessel. Hence, buyers know which boats they can trust to meet their quality standards. Especially when supply is short due to bad weather, or when large numbers of part-timers start flooding the market over the summer months with relatively cheap fish, buyers might need to acquire information on vessels they have not had any previous relations with. In other words, buyers take risks when buying fish from boats they have no previous experience with. Field observations have shown, however, that with the advent of digital vessel-tracking websites, buyers no longer rely only on previous transactions, but engage in real-time observations that provide valuable information that is neither based on past transactions, nor on relations of trust, nor is it captured by the formalised grading systems of the auction market. But what advantage can buyers gain by observing vessels on the internet?

In an interview about his use of the fish auction a processor explains that the quality of the fish depends strongly on the fishing grounds:

AD: But you don't get that information at the auction, where they fish?

Processor: No, but we know the boat's name...

AD: Aha.

Processor: And we can see it, we can see it! You know, boat's name: okay, he was there – we can see it! (MXIX)

The processor turns his computer screen towards me, where the website of marinetraffic.com displays current vessel activity around the Westfjords. He clicks on one of the red rectangles, which signifies a fishing vessel. Promptly, a small box displays a photo, name, course and speed of the vessel.

AD: So you try to get all this information?

Processor: Yeah, it's a lot of information, both the auction, and since we got the marinetraffic, it's, it's very good, and also just to see what is happening.

How buyers use information from vessel tracking technology for buying fish can be further illustrated by an example from a local fishmonger, who buys smaller quantities for his store and delivers to local restaurants and canteens. In his small office behind the counter, which only contains a few folders, a small desk and a PC displaying the interface of a vessel-tracking website, he explains to me how he evaluates from what boat to buy what type of fish on a particular day, based on the information displayed on the website. Accordingly, he tries constantly to keep track of vessel activities for three main reasons. First, to obtain general information about the availability of raw materials on a specific day. Second, to select which vessels might be of interest on a specific day. In this case, he might call a vessel of interest in order to obtain more information about the catch. Moreover, he can use his informational advantage to head off a vessel of interest directly at the harbour and use his temporal advantage to strike a deal directly with the fisher before the catch is put on auction, where he would risk losing out to another bidder. Most importantly, however, the information provided by the vessel-tracking site gives him information on the potential quality of a vessel's catch based on its fishing location. He explains this with a current example on the computer screen: clicking on one of the red rectangles located not far from a small island in the Ísfafjörðurdjúp, the biggest and deepest fjord of the Westfjords region, the fishmonger explains that some small boats tend to fish inside the fjord during bad weather, especially during the wintertime when Arctic blizzards make fishing on small boats all but impossible for weeks in the open waters of the Arctic ocean. Under normal circumstances, he would generally avoid buying fish from the fjord, which usually holds only smaller fish that are usually infested with parasites that come along with the large population of seals in the area. This, however, does not mean that he would not buy any fish from the boat displayed. In fact, the opposite is the case: especially haddock, a pelagic species feeding in open waters, is known to be of good value from these grounds, which are known to hold a reliable stock of high quality fish when that water is cold inside the fjord over the winter. He tells me, however, that in no circumstances would he buy ground fish such as cod from the boat, as these are known to be highly infested with parasites from these fishing grounds (FN: 36). This avoidance-strategy is confirmed by another buyer, who states that 'we can see [the boats, AD] on marinetraffic, we know the parasites area' (MXIX).

Enframing Local Knowledge

The field examples together show how vessel-tracking websites are used as market devices for coping with the problem of uncertainty. Technology, however, can never be understood as neutral or isolated from its social context, but must be seen as entrenched in the historically contingent social orders in which it is deployed. In line with Heidegger's (1977) account of technology as enframing (see Chapter 6), we can say that buyers' local knowledge and community ties are recontextualised and confronted with new pathways of practice based on the spatial and temporal synchronisation of market structures and fishing activities. In order to evaluate quality by means of vessel-tracking websites, therefore, it is not sufficient to rely only on the information displayed on the computer screen. Rather, one must draw on local knowledge such as 'knowing the parasites area' (MXIX) in order to engage in meaningful interpretations of the data. For instance, field interviews show that many buyers know the fishing grounds from their own experience as fishers: 'I've been to sea also, if I see the boats on the marinetraffic, I know how the fish are' (MXV). Fish stocks, however, are dynamic entities ('Yeah, it changes a lot', ibid.), so buyers try to maintain good ties with the fishers to keep up their local knowledge of the fishing grounds: 'But you keep track of that. It's important to talk a lot with people at the harbour' (ibid.).

To repeat the main argument: before the advent of scopic valuations, buyers' knowledge of fishing grounds was detached from fishing activities, and market making was based on the temporal and spatial separation of fishing and auctioning, leaving buyers to rely on nothing but their trust, local knowledge and network ties. Today, this local knowledge is recontextualised within digital frames that synchronise buyers with fishing activities that allows them to cope with the fundamental problem of quality uncertainty in spatially dispersed markets.

Controlling Market Identities

While discussing the issue of pricing and quality with Bjartur it becomes evident that he shows a high degree of reflexivity about his market identity and knows that they can achieve high auction prices only by maintaining high-quality standards over time:

> *AD*: But I guess if you're not directly involved in the processing, then you probably don't really care about quality?
>
> *Bjartur*: No, I think every fisherman in Iceland they want to come with the, to land with the good fish because if you come with good fish and sell it on the market, the person who is buying the fish, he know the names of boats, and if he sees it's a bad fish that has not been cut right away, not been put in cool water (...), maybe he says 'Okay, I'm not going to buy fish from this boat again' or maybe just put a lower price. If you come with good fish into land, then there's more possibility to get a higher price for the fish... (MII)

Furthermore, he seems to be well aware that potential buyers even observe his activities at sea in order to draw inferences on the potential value of his catch:

> Because now we have the 'eyes', you can see the boat on marinetraffic, and the fish buyers, they go on the computers, they see where the boats are, they read the names of the boats, they see this boat is very close to land and they didn't want the fish that was caught very close to the land because they have worms inside the fish. So they say 'Okay, we're not going to buy fish from these boats because they are very close to land' and they look out whether the boat was far out and say 'Okay, these boats have far better fish' – more outside, there is the good fish! And when the boats come and land, then the buyers know where they were fishing, and they buy from the boats that was more outside and with the bigger fish and they also know who was coming with fish, who has cooled it down, that was one part where they were fishing and one part was how do you, how do you say it - how do you take care of the fish on the boat. (MII)

Hence, observational scopes have not only opened up new ways of coping with quality uncertainty for buyers of fresh fish, but discipline the practices of quality production at sea. Although skippers have no information on whether and when they are being 'watched' by potential buyers, they know that they *might* be observed at any time. For this reason, fishers do not try only to meet the expectations of potential buyers for singular transactions, but engage in long-term control of their market identities (White, 2002, 2008), both analogue and virtual. It is therefore important for a vessel to be visible on vessel tracking websites, and uploading pictures of vessels can also be understood in terms of creating credibility in relation to market observers. Hence, scopic media not only mirror the current position of a vessel by means of direct a-presentation in a given synthetic situation (Knorr Cetina, 2009), but may transcend the present by imprinting a history of a vessel's positioning in a vessel's market identity. Buyers learn where boats tend and if they are reliable sources for daily transactions not only based on previous transactions, but also based on daily observations of their virtual identities in synthetic spaces of observation and control.

Conclusion: Towards a Panoptic Valuation Regime

This chapter has shown how AIS-technology has become established as market device that has transformed the temporal and spatial orientation of economic practices in the Icelandic coastal fisheries. Today, buyers on auction markets make routinised use of vessel tracking websites to reduce quality uncertainty by linking market activities with vessel positioning at sea. More recently, nobody else than the chairman of the National Association of Small Boat Owners (NASBO) himself has even suggested to institutionalise vessel tracking in the auction market by integrating AIS-gathered information on fishing grounds in order to allow buyers to trace their catch to the source (Fiskifréttir, 2017). As it seems, the organisation's anti-market rhetoric of the 1980s has given way to a new era of market-making and value generation in the name of quality.

In this new scopic valuation regime, remote monitoring of vessel activities is not limited to the auction market, but can also be found among processors with a vertically integrated fishing fleet. These vessels are widely decoupled from market transactions, as skippers work only on a contractual basis. This does not mean, however, that these fishers are under less pressure to put to sea. Often, quite the contrary is the case with regard to the high demand for raw materials in the processing plants. Hence, if a company's vessels cannot put to sea for a longer period of time, processors are forced to buy fish for comparatively high prices from the fish auction to fulfil their contracts with foreign buyers. Although processors or quota-owners can never force their skippers to put out to sea, they can put indirect pressure on them by comparing them with other boats:

> *AD*: If you talk to the owners, they tell you I can never demand or tell them that they have to go out, I mean how do you pressure people in this system?
>
> *Skipper*: Like in the neighboring village, when the guys are at home and the weather is bad, he [the owner of the processing plant] sometimes calls them to tell them that the boats from a village are at sea, so he is

calling them to let them know that the boats are at sea, he is automatically pressing!

AD: Aha, because he is comparing them with others…

Skipper: Yeah, and then they start to think. And sometimes he calls them too, to tell them to go to sea, [because he needs] fish for the factory … (MVII)

The rumours around this practice, however, are no secret, as the following quote from a plant owner makes clear:

No [you cannot force them], but you can ask them: why are you not at sea today with the others [laughter]? Na, we, hire a man to do that, you know, we hire the captain, and we really believe if he doesn't go out he will not get paid, so, that's very easy! He goes out if it's possible, and they do that! (MXIX)

Hence, the synthetic spaces of observation created by scopic media allow for a comparison of boats around the country and amplify the expectations and pressure put on fishers, who discipline their practices accordingly, as one skipper desperate about not having been able to fish in recent weeks due to bad weather reveals: 'I mean, he doesn't have to pressure me, I will do it myself, I will pressure myself. It has been like you say, maybe two or three weeks [of inactivity], I will try!' (MXX).

All in all, scopic media not only open up new pathways for more dynamic forms of economic valuation, but also of market control by creating feedback loops for fishers, who control their identities by disciplining their practices to the expectations of buyers and quota-owners, respectively. Hence, new digital spaces of observation add a new dimension to economisation by matching supply and demand through linking and synchronising fishing practices with production.

It seems as Bjartur, our independent small-boat fisher has not only become re-entangled in a global financial system (Chapter 3), but also into an all-encompassing *fishery panopticon* of mutual observation and control that inevitably reminds us of Foucault's (1975) famous analysis of Jeremy Bentham's prison utopia. Thus, instead of directly policing fishers at sea, the emergence of digital communication and tracking technologies has led to the emergence of a new, more subtle and dynamic form of surveillance and control in which fishers discipline their practices according to the rules and conventions of an omnipresent valuation regime.

The analysis of this new panoptic valuation regime completes our analysis of Bjartur's metamorphosis in liberal capitalism, in which the economic incentives of 'the market' have widely replaced the bureaucratic and redistributive top-down logic of the state. This general development can be seen in light of what Johnsen (2013) has diagnosed in line with Foucault (2009) as a general shift towards an increasing *governmentalisation* of the fishing industry in economic and juridical incentive structures that call for technologies of indirect surveillance and control. Hence, a new decentralised and dynamic management-regime of has not only optimised the flow of supply and demand by linking and synchronising fishing and processing, but also the self-enforcement of conventions and quality standards that lie at the heart of valorisation and profit-making (see Chapter 7).

While some might see this revaluation of coastal fisheries as producers of high quality goods as win-win situation that allows fishers, quota owners and processors to reap profits from an exclusive market niche, it nevertheless seems as the constant drive towards valorisation and profit-making has also become the industry's Achille's heel. Although the Icelandic fishing industry has certainly benefited from the low króna in the post-crisis era, six years later the value of the currency is again on course towards post-crisis levels. Hence, a reinvigorated króna plus increasing competition from the Barents Sea in which Norway and Russia have increased their annual fishing quotas has created much uncertainty among Bjartur and his fellows who even fear that quality orientation might even become a disadvantage if the pressure on prices continues to squeeze profits: 'We have to step back if we're going to sell the fish, we have to step back and get a lower price!' (MI).

How long middle-class consumers are willing to pay premiums to please their conscious remains to be seen. For Bjartur and his fellows, the struggle for independence will continue notwithstanding.

References

Akerlof, G. A. (1970). The Market for Lemons: Uncertainty and the Market Mechanism. *The Quarterly Journal of Economics, 83*(3), 488–500.

Bestor, T. C. (2004). *Tsukiji: The Fish Market at the Center of the World.* Berkely, CA: University of California Press.

Callon, M., Millo, Y., & Muniesa, F. (Eds.). (2007). *Market Devices.* Malden: Blackwell.

Dobeson, A. (2016). Scopic Valuations: How Digital Tracking Technologies Shape Economic Value. *Economy and Society, 45,* 454–478.

Dziewicki, M. (2007). *The Role of AIS for Small Ships Monitoring.* Paper presented at the Baltic Master Workshop, Gdynia, 11–12 May 2006.

European Commission. (2012). *Control Technologies: The EU System for Fisheries Controls.* Retrieved from http://ec.europa.eu/fisheries/cfp/control/technologies/index_en.htm.

Fiskifréttir (Producer). (2017, 11 September 2018). *Mikilvægt að forðast ormaslóð.* Retrieved from http://www.fiskifrettir.is/frettir/mikilvaegt-ad-fordast-ormaslod/138149/.

Fiskistofa. (2013, September 26). *AIS Transponders in Iceland.* Personal Communication with Thorstein Hilmarsson.

Foucault, M. (1975). *Discipline and Punish: The Birth of the Prison.* New York: Vintage.

Foucault, M. (2009). *The Birth of Biopolitics: Lectures at the Collège de France 1978–1979.* Basingstoke: Palgrave Macmillan.

Graham, I. (1999). *The Construction of Electronic Markets* (Ph.D. dissertation). University of Edinburgh, Edinburgh.

Heidegger, M. (1977). The Question Concerning Technology. In M. Heidegger (Ed.), *The Question Concerning Technology* (pp. 3–35). New York City: HarperCollins.

Johnsen, J. P. (2013). Is Fisheries Governance Possible? *Fish and Fisheries,* Online publication.

Knorr Cetina, K. (2003). From Pipes to Scopes: The Flow Architecture of Financial Markets. *Distinktion: Scandinavian Journal of Social Theory, 4*(2), 7–23.

Knorr Cetina, K. (2009). The Synthetic Situation: Interactionism for a Global World. *Symbolic Interaction, 32*(1), 61–87.

Knorr Cetina, K., & Bruegger, U. (2002). Global Microstructures: The Virtual Societies of Financial Markets. *American Journal of Sociology, 107*(4), 905–950.

Marinetraffic.com. (2013). *Frequently Asked Questions.* Retrieved from http:// www.marinetraffic.com/ais/de/faq.aspx?level1=160#2

Norris, A. (2007). AIS Implementation—Success or Failure? *The Journal of Navigation, 60*(1), 1–10.

Tetreault, B. J. (2005). *Use of the Automatic Identification System (AIS) for Maritime Domain Awareness (MDA).* Paper presented at the Oceans, Washington, DC.

White, H. C. (2002). *Markets from Networks: Socioeconomic Models of Production.* Princeton: Princeton University Press.

White, H. C. (2008). *Identity and Control: How Social Formations Emerge.* Princeton: Princeton University Press.

9

A New Culture of Liberal Rural Capitalism

> *Independence, and plenty of it. But where is their independence, may I ask?*
>
> Old Fritha, Halldór Laxness' *Independent People*

In *Independent People* (2008) Halldór Laxness writes:

> The history of centuries in this valley is the history of an independent man who grapples barehanded with a spectre which bears a new and ever a newer name. Sometimes the spectre is some half-divine fiend who lays a curse on his land. Sometimes it breaks his bones in the guise of a norn. Sometimes it destroys his croft in the form of a monster. And yet, always, to all eternity, it is the same spectre assailing the same man century after century. (ibid.: 17)

The independent man fighting against the spectre is our tragic hero Bjartur who defies the spirit of Kolumkilli, the ghost of an Irish sorcerer who laid a curse on the invaders coming from Norway. According to the scripture, Kolumkilli put a spell on the land of Bjartur, which has ever since been haunted by an evil woman named Gunnvör. Enlightened by the modern spirit, however, Bjartur 'the bright' believes in the idea of the autonomous

© The Author(s) 2019, corrected publication 2020
A. Dobeson, *Revaluing Coastal Fisheries*,
https://doi.org/10.1007/978-3-030-05087-0_9

subject as personified in the idea of the independent farmer, rather than in the myths of the old. For this reason, Bjartur refuses to cast a stone on Gunnvor's cairn on the mountain pass to his land to appease here ghost as is custom amongst superstitious people and re-names the farm from 'winterhouse' to the more sanguine name 'summerhouse' to symbolise the dawn of a new era after decades of exploitation as a peasant. Although it was tough withstanding the long and hard winters to come in a small and primitive farmhouse, Bjartur manages to pay of his debt for the land with thrift and his unbreakable work-ethos and stubbornness, which often is unbearable for his beloved ones. With the beginning of World War I, however, the farming business starts booming due to increasing demand and prices of meat and cotton production. With the general improvement of the situation of the farmers, Bjartur is even persuaded against his ideals by the local consumer-cooperative association to expand his farm by building a new large farmhouse with the help of new credit. After some prosperous years, however, demand for Icelandic meat and cotton starts declining again with the end of the war. Although Bjartur managed to finish building his new farmhouse, he is soon struggling to pay off the interest to the cooperative, leaving him to selling off his cow and some sheep. Soon, Bjartur remains empty handed and left behind alone and miserable without any means in a dull, cold and damp farmhouse, finally betraying his moral ideals by accepting help from others.

Although Bjartur of Summerhouses was a farmer, he bears strong parallels to our story of his fishing namesake. As Laxness' icon, the Bjartur of this story is determined to fulfil his childhood dream of rural independence. Both indebt themselves to be free and struggle likewise to harvest nature's resources under the auspices of their creditors in a rugged environment of long winters, rough weathers and the unpredictable cycles of boom and bust. As it seems, both seem to grabble with the same spectre that has haunted the land of Summerhouses over centuries. But is Bjartur of Summerhouses' cousin really doomed to loose the battle against Kolumkilli?

For allowing a careful assessment of this question, we have to better understand the nature of this battle, which will lead us to the question *who* Bjartur really is. The following section will therefore rephrase the question after Bjartur's identity and reconsider the question with regard to the narrative of our journey, which is summarised below.

Who Is Bjartur?

The romanticised ideal of the independent rural dweller as a primordially free and locally rooted farmer has a long tradition in modern literature. In particular, Knut Hamsun's anti-modern hero Isak, in his Nobel prize-winning novel *Growth of the Soil* (1917) embodies the image of good old farm life, which—in contrast to alienated city life with its unproductive work and ambiguous talk—helps him to find a primordial relationship to a harmonious nature in the rural north of Norway through hard but honest work.

In contrast, to Isak, however, Bjartur is continuously struggling with the misery and harshness of the Arctic environment: sheep die, ships sink, quotas get cut, prices change—businesses go bust. In other words, Laxness' observations of rural life have more in common with our observations than Hamsun's, as he does not conceptualise the independent rural dweller as anti-thesis to the rest of modern society. Rather, it seems as the opposite is the case. So is Bjartur a modern hero struggling against the structural constraints of rationalised modernity?

From a liberal perspective, Bjartur is a free and independent actor who is continuously defending his independence against the structural constraints of modern society, endlessly pending in the dialectics of *liberty* and *discipline* (Wagner, 1994). This perspective indicates that Bjartur is defending a once more or less harmonious community against the constraints of state bureaucracy and corporate control.

Our story, however, has questioned if this harmonious state has ever existed along the rugged coastlines of the island state, as the history of coastal communities as stable settlements seems to be tightly knit to the organisation of trade and markets.[1] Moreover, the story of emancipation takes the Kantian idea of a free rational subject as focal point of their analysis for granted and tell the story of its struggle and demise. Ways out of this dilemma are therefore usually located within the boundaries of the rational subject itself who defends her liberty by means of political mobilisation.

---

[1]See Dobeson (2019, forthcoming).

The story of Bjartur,—and in the end any relevant scientific observation is contextualised and moulded into a narrative—however, suggests yet another alternative interpretation, which neither romanticises a primordial form of rural life, nor reduces modern fisheries to a simple struggle of a more or less rational subject against the structural constraints of modern society. Rather, our journey into the world of fishing suggests a third perspective in which Bjartur is both agent and object (or: hero and victim) of socio-technical change and cultural transformation. This narrative, in line with Laxness monumental novel, deciphers Bjartur's self, that is his stubbornness, his striving for independence and community values, his skills at sea and ashore as a relational configuration that is grown out of societal discourse and practice. Hence, instead of telling the old tale of David fighting Goliath, Bjartur represents two sides of the same form—entanglement and disentanglement—, which gain their momentum in the sway of daily coping with a highly volatile and ever-changing environment. It is this pending between expectations and practices that will turn Bjartur against his own beliefs and cultural ideals.

While humans constantly create and transform their world by constantly engaging with their socio-technical environment and others, our journey into the world of market-based resource management has shown, however, that the frame of reference for daily coping has changed dramatically with the rise of neoliberal reform and the implementation of property rights-based management regimes. In order to fully understand the general consequences of this transformation for rural economies, this conclusive chapter will recapitulate the main observations around the guiding questions posed at the beginning of our journey: How can small and labour intensive industries navigate liberal capitalism? And what are the consequences of new markets and property right for rural development?

(Dis)Entangling Rural Economies

Rural economies are typically embedded in networks of production made up of social relations (Acheson, 1988; Barnes, 1954), institutions (Apostle et al., 1998; Hersoug, 2005; Hersoug, Holm, & Rånes,

2000; Holm, 1995) and discourses of production (Einarsson, 2011a; Helgason & Pálsson, 1997; Pálsson, 1991; Pálsson & Durrenberger, 1982). Most recently, accounts highlighting the discursive embeddedness of fisheries have emphasised the role of science and technology for the construction and organisation of market-based resource-management regimes (Holm, 2001; Holm & Nolde Nielsen, 2007; Johnsen, 2004; Johnsen, Holm, Sinclair, & Bavington, 2009). In order to understand how markets have reconfigured the dynamics of rural economies, however, we emphasised the practices of daily coping, in which these socio-technical arrangements are grounded and reproduced.

In a first step, we have seen how the organisation of markets for fishing rights and raw materials has reconfigured the role of the fisher from rural peasant to independent entrepreneur and investor: while in traditional rural production networks coastal fishers were bound to sell their catch to local processing plants, the organisation of markets for fishing rights and raw materials has *disentangled* coastal fishers from their community ties and emancipated them from the locally bound network of production by translating hierarchical and long-term trust-based relations into short-term oriented relations mediated by money-exchange. The more fishers started playing along with the market system, however, the more they were *re-entangled* in a new web of money-mediated relations with their creditors to guarantee financial liquidity to stay afloat. As a consequence, fishers and quota-owners had to increasingly re-orient their economic practices from cost-awareness towards profit-making, which opened up an endless cycle of investments in fishing quotas and increasing debt. Moreover, some quota-owners even saw the opportunity to speculate on rising quota prices until the financial bubble burst in 2008. As a consequence, the cultural ideal of coastal fisheries as backbone and symbol of independence has been challenged by practices of economising, which has re-valued small boats from being humble means and symbols of rural independence to highly valuable objects of investments and profit-making.

In a second step, we have seen that processes economisation and marketisation cannot be sufficiently understood in isolation from the economic domains that lie at the heart of value-creation: fishing and processing. The phenomenology of fishing has therefore shed some light on the practices and technologies that constitute the world of modern market-based fishing. In line with the early anthropological works of Heidegger's (1962) and Dreyfus (2014), ethnographic observations have shown that daily economic coping always already takes place in a highly routinised setting of a meaningful and pre-discovered relational environment of equipment and others. Moreover, we have seen that, despite the mechanisation of labour activities and the use of digital equipment, the swift coordination of activities and the spontaneity required for adjusting to the ever-changing and potentially dangerous environment of the sea requires skills that are acquired 'on the job' in an apprentice–master relation. Within this routinised coping, *circumspection* rather than intentional fixation on a certain device or event characterises the primary relation of the fisher with her environment. It is only in the case of failure—for instance, when a fishing line gets tangled—that the social order of routinised coping collapses and the fishers engage in reflexive activities that isolate and objectify their environment as means of causal problem-solving. In this context, the ethnographic material suggest that modern technology cannot be reduced to a means–end relation that increases the efficiency of the fishing operation, but must be understood in line with Heidegger (1977) as *enframing* of daily practices, which is henceforth directed to the objectification and ordering of 'nature' as a resource that can be extracted and stored over time. In addition to Heidegger's analysis of 'analogue' technology, the case shows how digital technologies intensify this enframing of the marine environment by not only synchronising fishers with their competitors at sea, but also with the fish that can be tracked and observed in real time.

In modern fishing, however, the contingency of an ever-changing environment not only interferes with practices at sea, but is intertwined with the domestic world of changing rules, regulations, stock predictions, banks and multiple market structures, as the example of the 'haddock crisis' has shown. Thus, processes of economisation do not simply imply rationalisation of operations towards economic efficiency, but point towards flexible adaption through practical coping to an ever-changing contingent environment.

In a similar way, traditional capture technologies are not simply replaced by more efficient ones, but recontextualised within their current socio-technical environment. Hence, changing market structures and new technologies in fisheries and processing have come along with a transformation of practices and a *revaluation* of the identity of the coastal fisheries as suppliers of a high-quality market niche for 'line-caught fish' that is materialised in the socio-material practices of fishing and processing. In contrast to durable goods, however, easily perishable raw materials such as fish require special treatment for maintaining high quality and value. Thus, due to the increasing economisation of the coastal fisheries, fishers adjust their practices from a quantity- to a quality-orientation in order to achieve the highest possible prices at auction and create long-term ties with buyers engaging in the marketing of 'sustainable' and 'line-caught products'. Thus, the construction of a top-quality market niche is not the result of individual entrepreneurship, but must be understood in light of a broader collective transformation of economic practices that respond to neoliberal discourses of sustainability and green consumerism.

Within this transformation, a new 'scopic' valuation regime reduces quality uncertainty by linking and synchronising buyers with fishing activities at sea. Hence, digital information technologies such as AIS and scopic media such as chart plotters and computer screens are not only used by skippers for observing each other, but also used as evaluative devices by market actors. Hence, although network ties still play an important role for the generation of trust between fishers and buyers, scopic media allow experienced buyers informational advantage on the auction market and the remote evaluation of quality based on a vessel's positioning at sea. Furthermore, quota-owners use the information provided for controlling their vessels by comparing them with the activities of their peers at sea. Thus, digital technologies likewise enframe fish and fishers within an all-encompassing fishery panopticon of mutual surveillance and control.

All in all, the case of the Icelandic coastal fisheries shows that traditional forms of production have not been simply replaced by economically more 'efficient' alternatives. Rather, existing cultural and socio-technical pathways have been revalued and adjusted within the context of new markets and technologies that have opened new pathways for economic practices. Thus, economisation is neither a one-sided universal structural force that has simply infiltrated coastal communities, nor simply a conglomerate of individual rational choices. Rather, it is a translation of socio-technical arrangements, institutions, identities and network ties that has been unfolded in an oscillating process of disentanglement and re-entanglement of economic expectations that transforms the culture of the coastal fisheries from symbol and means of rural independence to a new liberal culture of market-based entrepreneurship and profit-making.

Above all, the our journey has shed some light on the widely neglected field of natural resources in economic sociology. By doing so, it adds new empirical knowledge to our understanding of the dynamics of economisation and marketisation in rural production networks. What is different from previous accounts is that processes of economisation are conceptualised in the contingent situatedness of economic practices that follow and modify pathways in which the small pragmatic adjustments of daily economic coping take place. Hence, the study has shown that the 'primacy of the economy' in modern society (Beckert, 2009) does not contradict the survival of small artisanal industries depending on their embeddedness in networks, institutions and socio-material discourses of production. The economisation of the fisheries economy by markets and property rights has furthermore shown that gradual pragmatic coping and changing practices do not only reconfigure institutions and network ties, but also transform the culture of coastal fisheries and rural economies itself (see below).

Second, observations of a fishing vessel reveal that modern technologies do not simply determine human actions in the sense of a 'harvest machinery' (Johnsen, 2004) that simply replaces traditional forms of knowledge, technology and skill. Rather, ethnographic observations indicate that new technologies gain their meaning and momentum only in relation to their local epistemic cultures of production. Hence, only within the contingent situatedness of daily skilful coping do new technologies gain their meanings and may open up new pathways for practices in which traditional knowledge is recontextualised.

Third, we have decentralised and broadened our understanding of risk and uncertainty in the economy (Beckert, 1996; Knight, 1921) by highlighting the role of the more-than-human environments of markets and production. Accordingly, fishers not only rely on cognitive frames and coping devices to reduce economic risks in markets, but deploy socio-technical coping strategies that make the market fit with the highly volatile and ever-changing flows of the sea. Furthermore, we have acquired more comprehensive understanding of the role of technology for the dynamics of contemporary capitalism than the literature on market devices suggests (Callon, Millo, & Muniesa, 2007). Accordingly, daily economic coping with the environment unfolds the 'paradox of technology', which on the one hand solves economic problems by means of functional simplification, which tend to be undermined by the complexities of an ever-changing environment, consequently triggering an endless cycle of new investments and breakdowns.

Fourth, our journey has contributed to sociology valuation (Antal, Hutter, & Stark, 2015; Beckert & Aspers, 2011; Beckert & Musselin, 2013; Lamont, 2012), which has mainly treated economic valuation independently of the materiality and practices of production. The case of the Icelandic coastal fisheries shows clearly that economic value, in particular in natural resource-based economies, cannot be understood as detached from the practices that lie at the heart of value creation in rural networks of production, as they are grounded and materialised in the fundamental practices of fishing and processing. The same holds true for the construction of 'quality' (Beckert & Musselin, 2013), which has to be seen as the result of a collectively negotiated process that is shaped and materialised within the epistemic cultures (Knorr Cetina, 1999) in which daily economic coping takes place. Hence, the changing practices towards quality-upgrading have to be understood in response to the increasing economisation of the Icelandic fishery economy, in which traditional forms of knowledge on processing and fishing are reinterpreted in order to adapt to a changing economic environment.

Finally, we have witnessed how new technologies such as AIS open up new pathways for practices that reconfigure the temporal and spatial configurations of rural production networks in which exchange of information and resources used to be based on lagged distribution through interpersonal network ties (Acheson, 1988; Bestor, 2004; White, 1981, 2002). Hence, with increasing media institutionalisation of the economy, buyers and sellers today are increasingly connected independently from their network ties in synthetic spaces of observation and control, which link and synchronise activities based on remote real-time observations. This insight is not only of major importance for understanding of how economic actors cope with quality uncertainty in markets, but also to understand the socio-technical rationale of an increasingly decentralised and all-encompassing fishery panopticon.

We have now provided a comprehensive explanation of *how* markets and property rights reconfigure relations and practices in rural networks of production. In what follows we will dig deeper into the meaning of this transformation of rural economies in the wake of new markets, property rights and technology, which is materialised in a new *liberal* culture of rural capitalism.

The Transformation of Rural Independence

From a resource optimisation perspective, the economisation and marketisation of the Icelandic small-boat fisheries appears to be a win-win situation for all participants. Those who were too old or inefficient were reimbursed for leaving the industry while those remaining contributed to an overall valorisation of the small boat fisheries by going from quantity to quality-oriented niche marketing. The development of the small-boat fisheries in the wake of new markets and the ITQ-system in general and its de facto closure to newcomers, however, has sparked fierce controversies about the concentration of ownership rights and the closure of the marine commons as fundamental right of the Icelandic people (Einarsson, 2011b). Reforming the system from within, however, has proven difficult as family run businesses have been turned into petit capitalist enterprises who's survival is dependent on the volatile world of international capital markets. It would however be premature to see individual motives such as 'greed' and one-sided profit-orientation as sole motivator behind this development. The real issue at stake for rural development behind the conflicting values and individual dilemmas is that small-boats have lost their initial function by providing flexibility to coastal communities to counterbalance the risk and uncertainties typically associated with capital intensive economies of scale. Thus, the economisation of the small-boat fisheries has furthered the economisation of the rural with the rise of a new culture of liberal rural capitalism in which private ownership structures, individual entrepreneurship and market performance decide who stays afloat, rather than collective belonging, community-based forms of solidarity and redistribution. As a consequence, coastal economies are no longer based on the principle of solidarity as represented in the culture of the independent small-boat owner who owns *and* fishes their quota and contributes with the catch to the local economy, but justified alone by 'the market'. Standing in the shadow of their creditors while being entangled in international credit and consumer markets, however, quota owners can decide from one day to another to move fishing vessels and landing ports to cut costs, leaving little work for the people in peripheral coastal communities.

Empirical evidence of this de-solidarisation can be found not only in the prioritisation of economic ends over the needs of the community and the development of (electronic) control regimes, but also in the political fragmentation of the representative organs, such as the National Association of Small Boat Owners (NASBO), as larger quota-owners have founded their own lobbying organisations to communicate their economic interests in the political domain. One of these interests is the abolition of the subsidies for hand-baited lines, which would put many people in communities depending on coastal fisheries out of work. Another is the political mobilisation of large quota-owners for bigger vessels in the coastal fisheries.[2]

In the end, the increasing fragmentation and instability of the coastal fisheries not only create problems for the fisheries as such, but also for the service people, whose jobs largely depend on the wealth created in the fishing sector. Evidence for this can be found in the rural real estate market, in which market prices correlate highly with the accumulation of fishing rights in the communities (Benediktsson & Karlsdóttir, 2011). Quota-owners can decide to sell off their quotas for any reason, be it due to financial speculation, miscalculation or simply

---

[2]At the time of writing (2014), the Icelandic government had doubled the maximum weight for coastal fishing vessels from 15 to 30 tonnes.

because of unforeseeable government decisions or market developments. Of course, booms and busts lie in the nature of capitalist economies, although in this case the initial function of coastal fisheries as solidaristic safety net is undermined by the logic of capital concentration: the more fishing quota are concentrated in the hands of a few owners, the more a fishing community will be vulnerable to economic decline in case of bankruptcy.

How Big Can Small Be?

Our journey into the world of contemporary coastal fisheries has revealed that social scientists searching for alternatives to market-based solutions and industrial rural capitalism can no longer simply assume that 'small is beautiful' (Schumacher, 1973), as 'small' and 'sustainable' have become itself an important feature of value creation that caters wealthy consumers' search for authenticity in liberal capitalism. With regard to the increasing social scientific interest in the role of small-scale fisheries as means of rural resilience and poverty alleviation (Chuenpagdee, 2011), this book has moreover given a more nuanced view of about the double-edged consequences of market-based management as so-called 'best practice solution' for rural development. The case has shown that, once implemented, property rights create robust global networks that mesh fishers and fish into a complex web of money-mediated relations, while at the same time transforming them into agents who themselves drive economisation in their daily copings. As the result, the boundaries of the coastal fisheries have not only been stretched far out in the Arctic Ocean, but also in the world of large-scale capital accumulation. All in all, the successive economisation of small boats has indeed turned a century old 'inefficient' tradition into a profitable business that competes indirectly with its large-scale industrial counterpart by establishing and maintaining a quality-oriented market-niche. But at what cost?

In the end, coastal fisheries, as well as our archetype of the independent small boat fisher are a cultural forms that have grown and survived the process of modernisation in the sway of political discourses and economic practices. The question therefore should not be whether one is for or against the 'free market', but about the boundaries, ownership structures and the regulations that frame and define the cultural form of the coastal fisheries in a market economy. The people of Iceland and of other fishing nations flirting with market-based solutions to resource management may ask themselves the following questions: What are the cultural and social values of coastal fisheries? How shall we define their boundaries? What is their function in contemporary society? Furthermore, should a few quota-owners have the power to accumulate large amounts of fishing quotas and vessels? Or even more critically, should not those who go to sea own the right to fish, as the old Icelandic proverb *Þeir fiska sem róa*—(Those fish who go to sea) suggests?

These questions ultimately come down to the most fundamental question, which bridges political discourse and academic sociology: in what kind of society do we want to live? Do we want to live in a hyper-urbanised society in which access to resources are owned few large companies who reap the benefits of an increasingly efficient and technisised sector?[3] Or do we want to live in a society in which people have the right to choose where to live in a thriving rural landscape that is more than a mere facade for the booming tourism sector?

Our journey into the world of market-based fisheries has now come to an end. This does not mean, however, that the research on the transformative power of markets, property rights and new technology has been exhausted. While our journey into the world of market-based fisheries allowed us to gain first-hand insights into the role of new markets, property rights and technology in the fishing industry, it goes without saying that more in-depth studies of economisation and marketisation of different rural economies are needed in order to fully understand the transformation of the countryside in the neoliberal age.

---

[3]This is of course not necessarily an argument against large-scale industrialism. The case of the Icelandic fisheries even suggests that the large-scale industrial fleet forms the backbone for stable markets and prices. From this perspective, large and small corporations represent a successful synthesis, serving different market niches.

References

Acheson, J. M. (1988). *The Lobster Gangs of Maine*. Hanover: University Press of New England.

Antal, A. B., Hutter, M., & Stark, D. (Eds.). (2015). *Moments of Valuation*. Oxford: Oxford University Press.

Apostle, R., Barret, G., Holm, P., Jentoft, S., Mazany, L., McCay, B., & Mikaelsen, K. (1998). *Community, State and Market on the North Atlantic Rim: Challenges to Modernity in the Fisheries*. Toronto: University of Toronto Press.

Barnes, J. A. (1954). Class and Committees in a Norwegian Island Parish. *Human Relations, 7*, 39–58.

Beckert, J. (1996). What Is Sociological About Economic Sociology? Uncertainty and the Embeddedness of Economic Action. *Theory and Society, 25*, 803–840.

Beckert, J. (2009). Wirtschaftssoziologie als Gesellschaftsstheorie. *Zeitschrift für Soziologie, 38*(3), 182–197.

Beckert, J., & Aspers, P. (Eds.). (2011). *The Worth of Goods: Valuation & Pricing in the Economy*. Oxford: Oxford University Press.

Beckert, J., & Musselin, C. (Eds.). (2013). *Constructing Quality: The Classification of Goods in Markets*. Oxford: Oxford University Press.

Benediktsson, K., & Karlsdóttir, A. (2011). Iceland: Crisis and Regional Development—Thanks for All the Fish? *European Urban and Regional Studies, 18*(2), 228–235.

Bestor, T. C. (2004). *Tsukiji: The Fish Market at the Center of the World*. Berkeley: University of California Press.

Callon, M., Millo, Y., & Muniesa, F. (Eds.). (2007). *Market Devices*. Malden: Blackwell.

Chuenpagdee, R. (Ed.). (2011). *World Small-Scale Fisheries: Contemporary Visions*. Delft: Eburon.

Dobeson, A. (2019, forthcoming). Das Fischerdorf im liberalen Kapitalismus: sozialräumliche Öffnungs- und Schließungsprozesse in der nordatlantischen Peripherie. In A. Steinführer, L. Laschewski, T. Mölders, & R. Siebert (Eds.), *Das Dorf. Soziale Prozesse und räumliche Arrangements*. Berlin: LIT.

Dreyfus, H. L. (2014). *Skillful Coping: Essays on the Phenomenology of Everyday Perception and Action*. Oxford: Oxford University Press.

Einarsson, N. (2011a). *Culture, Conflict and Crises in the Icelandic Fisheries: An Anthropological Study of People, Policy and Marine Resources in the North Atlantic Arctic* (PhD dissertation). Uppsala University.

Einarsson, N. (2011b). Fisheries Governance and Social Discourse in Post-crisis Iceland: Responses to the UN Human Rights Committee's Views in Case 1306/ 2004. In N. Einarsson (Ed.), *Culture, Conflict and Crises in the Icelandic Fisheries: An Anthropological Study of People, Policy and Marine Resources in the North Atlantic Arctic.* Uppsala: Acta Universitatis Upsaliensis.

Heidegger, M. (1962). *Being and Time.* New York City: HarperCollins.

Heidegger, M. (1977). The Question Concerning Technology. In M. Heidegger (Ed.), *The Question Concerning Technology* (pp. 3–35). New York City: HarperCollins.

Helgason, A., & Pálsson, G. (1997). Contested Commodities: The Moral Landscape of Modernist Regimes. *The Journal of the Royal Anthropological Institute, 3*(3), 451–471.

Hersoug, B. (2005). *Closing the Commons: Norwegian Fisheries from Open Access to Private Property.* Delft: Eburon.

Hersoug, B., Holm, P., & Rånes, S. A. (2000). The Missing T. Path Dependency Within an Individual Vessel Quota System—The Case of the Norwegian Cod Fisheries. *Marine Policy, 24,* 319–330.

Holm, P. (1995). The Dynamics of Institutionalization: The Transformation Process in Norwegian Fisheries. *Administrative Science Quarterly, 40*(3), 398–422.

Holm, P. (2001). *The Invisible Revolution: The Construction of Institutional Change in the Fisheries* (PhD dissertation). University of Tromsø, Tromsø.

Holm, P., & Nolde Nielsen, K. (2007). Framing Fish, Making Markets: The Construction of Individual Transferable Quotas. In Y. Millo, M. Callon, & F. Muniesa (Eds.), *Market Devices* (Vol. 55, pp. 173–195). Malden: Blackwell.

Johnsen, J. P. (2004). The Evolution of the "Harvest Machinery": Why Capture Capacity Has Continued to Expand in Norwegian Fisheries. *Marine Policy, 29,* 481–493.

Johnsen, J. P., Holm, P., Sinclair, P., & Bavington, D. (2009). The Cyborgization of the Fisheries: On Attempts to Make Fisheries Management Possible. *Mast, 7*(2), 9–34.

Knight, F. H. (1921). *Risk, Uncertainty and Profit.* Boston: Houghton Mifflin.

Knorr Cetina, K. (1999). *Epistemic Cultures: How the Sciences Make Knowledge.* Cambridge, MA: Harvard University Press.

Lamont, M. (2012). Toward a Comparative Sociology of Valuation and Evaluation. *Annual Review of Sociology, 38,* 201–221.

Laxness, H. (2008). *Independent People.* London: Vintage.

Pálsson, G. (1991). *Coastal Economies, Cultural Accounts. Human Ecology and Icealandic Discourse*. Manchester: Manchester University Press.

Pálsson, G., & Durrenberger, E. P. (1982). To Dream of Fish: The Causes of Icelandic Skipper's Fishing Success. *Journal of Anthropological Research, 38*(2), 227–242.

Schumacher, E. F. (1973). *Small Is Beautiful: A Study of Economics as If People Mattered*. London: Blond & Briggs.

Wagner, P. (1994). *A Sociology of Modernity: Liberty and Discipline*. London: Routledge.

White, H. C. (1981). Where Do Markets Come From? *The American Journal of Sociology, 87*(3), 517–547.

White, H. C. (2002). *Markets from Networks: Socioeconomic Models of Production*. Princeton: Princeton University Press.

Correction to: Revaluing Coastal Fisheries

Correction to:
A. Dobeson, *Revaluing Coastal Fisheries*,
https://doi.org/10.1007/978-3-030-05087-0

The book was inadvertently published with errors and the same has been updated later as follows:

Preface
p. v
'P.h.D. student' has been changed to 'PhD student'
p. viii
'Siavash Alimadadi': One of the two repetitions has been deleted

Chapter 2
p. 35
'Hence, the next two sections will present some perspectives that attempt to…' has been modified with the deletion of 'Hence' and start sentence as 'The next two sections…'

The updated version of the book can be found at
https://doi.org/10.1007/978-3-030-05087-0

Chapter 3
p. 52
'From this perspective' has been changed to 'Hence' in the sentence 'From this perspective, economisation is not merely a structural force of society' because 'From this perspective' already appears in the same section as start of a sentence)
p. 60
'It is however clear that the possibility of becoming a quota owner' has been changed to 'But it is clear that the possibility of becoming a quota owner', to avoid double use of 'however' in the following sentence
p. 60
The phrase 'of what it meant to be independent' has been changed to 'of what it means to be independent' in the sentence 'Rather, they were now confronted with a complex set of new rules and regulations, which changed the expectations of what it means to be independent'
p. 70
'It soon becomes clear that investments…' has been corrected in tense: 'It soon became clear…'
p. 81
Misprint has been corrected in the sentence 'These companies, however, are tied to their financial liabilities and are literally forced to fish off their debts and squeeze as much surplus value as possible of their quota shares…'
p. 84
'In this system, however, fishers cannot choose when to put to sea and therefore often put to sea in bad weather…'
- 'put to sea' appears twice in same sentence

Chapter 4
p. 87
Footnote 1 is deleted

Chapter 5
p. 135
The term 'movies' has been changed to 'films' (British English) in the sentence: 'Like the skipper on the way out, he will entertain himself most of the time by watching films'

Chapter 6
p. 143
The sentence 'The higher the financial stakes are in the industry, scientists will therefore have an endless incentive to translate uncertainties into risks by optimising their models and predictions to match the flows of the sea' has been changed as 'The higher the financial stakes are in the industry, the more scientists will have an incentive to optimise their models and predictions to match the flows of the sea'
p. 143
Footnote 11 is deleted as the quotation appears on the same page in main text
p. 151
The phrase '...inevitably produce socio-technical drawbacks in other domains of economic coping' has been reformulated as 'inevitably produce drawbacks in other domains'
p. 153
In Image 6.3, reference '(source: marinetraffic.com)' in caption is deleted
p. 154
The repeated phrase 'and potentially dangerous' has been deleted in the sentence 'Obviously, fishing always takes place in a potentially rough and dangerous environment'
p. 160
The sentence 'This paradox of technology - to loosely follow the thoughts of Luhmann (1993) on the relation between risk and technology amplified in an increasingly economised environment, in which technology is more and more used as means to control economic risks' has been amended to 'This 'paradox of technology—to loosely follow the thoughts of Luhmann (1993)—is amplified in an increasingly economised environment, in which technology is used as means to control economic risks'

Chapter 7
p. 179
The phrase 'as a processor working exclusively with small boats makes clear...' is amended to 'as a processor working exclusively with small boats explains...'

p. 186
The phrase 'Hence, in today's fisheries, new ways handling fish...' is corrected to: 'Hence, in today's fisheries, new ways of handling fish'

Chapter 9
p. 213
Word choice correction of changing 'go bankrupt' to 'go bust' is made in: sheep die, ships sink, quotas get cut, prices change—businesses go bust'

p. 213
In Footnote 1, amend to 'See Dobeson 2019

p. 223
The sentence 'Or even more critically, should not those who go to sea own the right to fish, as the old Icelandic...' is amended to 'Or even more critically, should not those who go to sea own the right to fish, as the old Icelandic proverb...'

Index
p. 265
- Entry is deleted for 'small'
p. 266
The sub-entry 'digital 19' under 'technology' is deleted

Appendix

Hook and Line Fisheries

This appendix gives the unfamiliar reader a chance to become acquainted with the basic capture technologies that have been the material backbone of the Icelandic coastal fisheries for centuries.

Jigging

Jigging (Icelandic *Handfæri*, literally meaning 'handline') is among the oldest fishing techniques. Jigs are a kind of artificial fishing lure that is typically made out of rubber that is attached to a hook. Most commonly, plain but flashy colours, such as yellow, red, white, green and blue are tied to a monofilament that is weighted down with a sinker. To increase capture efficiency, several—on average, 8–10—jigs are tied in a row to a so-called paternoster-rig. Depending on the number of jigs used, paternoster-rigs allow the catching of several fish in one turn. During fishing activity, jigs are jerked up and down to imitate living bait to enhance their attractiveness to the fish.

© The Editor(s) (if applicable) and The Author(s) 2019, corrected publication 2020
A. Dobeson, *Revaluing Coastal Fisheries*,
https://doi.org/10.1007/978-3-030-05087-0

Traditionally, one or more baited hooks and a sinker are tied to a long line. When the vessel—in the olden days an open rowing boat—has reached the fishing grounds and the foreman has given the signal to start fishing, the hook and sinker are let down to the desired depth by a single fisher, who controls the line with his hands. In a way, modern sea angling, which is practised for recreational purposes, resembles this traditional technique in its basic form, with the only difference that the angler uses a fishing rod and a reel for line control. In both cases, the fisher is dependent on his own perception that is mediated through the fishing line, which tells him whether a fish has taken the bait so that he can pull up and reel in the line.

In contemporary coastal fisheries, jigging still plays an integral part in the summer fishing season and is typically used on smaller vessels of around 5–10 tonnes. Although the old denomination *handfæri* is still used for classification in Iceland, technological innovations, in particular jigging computers, have increased efficiency[1] tremendously (see Image A.1). Jigging computers not only make it possible to pull up the catch by machine power, but also to programme the desired depth, among other functions. In the latest models, electronic sensors in the sinker provide information that is displayed on the computer screen. For instance, if a skipper sees a shoal of cod at a certain depth on the plotter of his fish finder, he can rapidly drop the bait down to the desired depth and gain information about the weight of fish already hooked. Thus, the uncertainty when reeling in the line with manual jigging machines is reduced and allows the much more efficient coordination of fishing turns. Furthermore, only one fisher—often the skipper of the vessel—is needed to unhook the fish while the other jigging machines are operating. Hence, jigging vessels are often controlled by only one, at most two fishers.

In Iceland, commercial jigging fishing is practised almost exclusively from May to August when the sea is warmer and the cod and saithe are actively feeding on big shoals of capelin, herring, mackerel, shrimps and other bait close to the coastal regions. According to the fishers interviewed, the fish do not take the jigs in the winter as they lie more passively on the sea bottom, waiting for

---

[1] Other ground species may be caught as by-catch. For some reason, the haddock, which is caught together with the cod, saithe and wolfish on the baited longlines, is seldom caught on handlines.

Image A.1 Jigging computer. Fully computerised Jigging machine (Photo by AD)

something to feed on to pass. Furthermore, jigging is dependent on relatively good weather with slow winds and smaller waves to ensure optimal operation of the jigging computers during a drift. If the drift is too strong, however, the vessel may be slowed down with a drift anchor to allow maximum performance of the jigging computers. When drifting, the skipper not only saves fuel, but also makes sure that the roaring engine does not spook the fish.

All in all, jigging is considered to be especially economic and environmentally friendly as it—unlike bottom trawling—protects the seabed when fishing for demersal species, minimises by-catches and saves (in comparison with longline fishing) a lot of oil.

Longlining

When the short Arctic summer passes and the seas get rougher, slightly bigger vessels fishing with baited long lines for more passive fish in colder waters take over. Longlining is a traditional capture technology that has been of particular importance for both past and current developments of the small-scale fisheries in the Westfjords. As the name indicates, a longline is a comparatively long fishing line equipped with perhaps hundreds of baited hooks that is put to sea to the stern while the vessel is slowly moving forward. After the line is set in the ocean, it is marked with a buoy and left in the sea until it is hauled aboard again to manually unhook the catch.

According to Sævaldsson and Valtýsson (2013), the first usage of longlines can historically be dated back to at least the late fifteenth century. Even though longlines were much more efficient than the manual handlines from which they originated, operating them remained quite expensive due to the increasing cost of labour and fresh bait, which was not always available and storable in large quantities over time. In addition, it was very difficult to control the longlines from open rowing boats in the strong currents and drift of the open ocean as lines would get snagged due to the drift when 'lying' on the sea bottom. Thus, a combination of these factors made longline fishing too costly for most boat owners as the risk of losing costly lines was high (Þór, 2013). The usage of longlines became more common in the nineteenth century, when fishing from sailing boats was booming, but it was not until the early twentieth century that two major technological breakthroughs laid the foundations for the triumphal march of the longline for the socio-economic development of twentieth century fisheries in Iceland. *First*, the invention of reliable freezing technologies, which not only revolutionised the processing industry and consumer markets in Western countries, but also allowed the storage of bait over longer periods of time; and *second*, the advent of motorised vessels, which gradually replaced sailing and rowing boats in the commercial fisheries and hence facilitated recapturing lost lines in the sea. Ever since, longlines have been of major importance for smaller decked vessels, particularly during the rise of the trawler industry, and could even outnumber them in catches for longer periods until the usage of gillnets during the spawning season grew popular in the 1960s (Sævaldsson & Valtýsson, 2013).

In the Westfjords region, however, the longline was able to establish itself much earlier as primary capture technology (Þór, 2002: 236). This socio-economic *Sonderweg* in the development of the fisheries in the Westfjords can be explained mainly by the geographical features of the region: in comparison with the unsheltered fishing grounds, in particular in the south–west, where most of the fishing activity was taking place, the rich fishing grounds in the Westfjords region were naturally sheltered from the strong currents of the open ocean by the numerous fjords in which they were located. This peculiarity of the region made it technically and economically viable to control the longlines from the open rowing boats putting to sea from the fishing stations inside the fjords—even during the wintertime, when seas usually get rougher.

Besides the main target species of cod and haddock, longlines in Iceland have also been used for capturing Greenland shark, whose liver oil used to play an important export role during industrialisation, in particular for illuminating street lights in the Continent's urban centres in the nineteenth century (Þór, 2002: 239). Especially from the fishing stations in the Westfjords region open rowing boats would sail offshore to bait with seals or other large and bloody pieces of meat for the northernmost shark species. Today, however, only a few vessels seasonally bait for Greenland shark as the demand for producing *Hákarl*,[2] a traditional fermented fish dish, is covered by the by-catches of the trawler fisheries.

Today, longlining is one of the traditional capture techniques that have survived both the revolution in the freezing trawler industry and the crises of the Icelandic fisheries in the twentieth century. Moreover, it has remained the most efficient and widespread technique in the small-boat sector for vessels under 15 BT (now 30 BT) and has proven to be of particular importance for the small-boat revival in some prospering villages in the Westfjords.

A modern longlining vessel's crew usually consists of at least two fishers, one of whom is usually the skipper and the other a deckhand who helps the skipper with setting and retrieving the line, unhooking and

---

[2] The flesh of the Greenland Shark is poisonous as it contains neurotoxins, which can be seriously harmful if consumed raw. The toxins can be decomposed by fermentation of the shark's meat. For this, parts of the shark are first buried for a few weeks and then hung up in wooden huts to let them dry in the cold Artic winds. In Iceland, Hákarl is traditionally eaten on special occasions like national holidays, in particular at *Þorrablót* in mid-February, during which the shark's meat is typically consumed and washed down with some Icelandic snaps (Brennivín), but it is usually available all year around and popular for offering to sceptical foreigners.

cutting the catch to let it bleed dry to ensure the best quality and highest prices. On larger vessels up to 15 BT, the crew can consist of 3–4 fishers, one of whom is usually the skipper, another an engineer and one or two deckhands.

Whereas in the olden days seasonal longlining was usually practised in the sheltered fjords during the winter months, today most fishing operations take place all year round on the open ocean as the quality of the fish, in particular cod, is believed to be of higher quality. Thus, fishing inside the fjords is commonly seen as a stopgap solution when the weather turns too rough, but the economic incentive for maintaining the fishing operation is high.

Like jigging, the longline fisheries have undergone a period of rationalisation, fostered by capital-intensive investments and technological innovations, which mainly resulted in comparatively cheap, plastic-hulled and fast decked vessels, which were capable of going out in rough weather, such as the *Cleopatra*-type fishing vessels build by the Icelandic company *Trefjar* from Hafnarfjörður (see Image 3.2) Thus, the latest developments in the industry allowed a revitalisation of the longline fisheries, which, unlike the seasonal jigging fisheries, gave investors the chance to invest in a year-round business harvesting fish and securing a more or less constant supply of fish for the local processing facilities.

Today, longlines are almost exclusively used for demersal fisheries. Cod and haddock are the main species caught, with catches lowest in the summer. However, also other valuable species, in particular wolfish in the spring, monkfish, tusk and ling are valuable by-catches (Sævaldsson & Valtýsson, 2013).

Artisanal Longlining as a Means of Rural Development

Today, artisanal baiting is subsidised by the government as a means of rural development. Accordingly, quota-owners can land an excess of 20% if lines are baited ashore, which most quota-owners do instead of using automated baiting machines. Lines are usually baited in special baiting houses or shacks, which are typically directly situated alongside the homeports of the vessels. This allows spontaneous communication and coordination between fishing crews and baiters, and transport distances are short. In general, baiting is low-skilled work that does not require any training and is typically done by teenagers, migrant workers or older people such as retired fishers who are paid per *bala*, the native term for the buckets in which the longlines are stored (see Image A.2). In the Icelandic demersal longline fisheries, each bala contains a fishing line of 500 metres equipped with 500 hooks at one metre intervals. Especially for the younger generation, working in a baiting house is an important social space where they can establish contacts with the fishers and dream of being offered a position as deckhand on one of the highly paid and respected fishing vessels.

Gill Nets

Another important technique for more capital-intensive vessels without a hook & line licence in the small-boat ITQ system are gillnets (Sævaldsson

Image A.2 Baiting the line. A worker baiting a line in the baiting house in a small fishing village in the Westfjords. After a chunk of squid or fish is attached to the hook, the baiter has to make sure to lay it into the *bala* in an orderly fashion to ensure that the line can be released smoothly during the fishing operation without getting tangled (Photo by AD)

& Valtýsson, 2012). These nets, each about 50m long, are put out on the sea bottom like a flat tent, which is kept vertical through buoys on the top. A big entrance is left for the fish to swim inside and get entangled in the mesh with their gills. There are different mesh-sizes for different kinds of fish. The main target fish is cod, but also saithe, while haddock, ling and monkfish are caught in lesser amounts. Other techniques such as purse seines are—in contrast to Norwegian regulations—generally prohibited in the Icelandic cod fishery, but used by large trawlers fishing for capelin, an important commercial fish for fishmeal and fish oil processing.

Figures

See Figs. A.1, A.2, A.3, A.4, A.5, A.6, A.7, A.8, A.9, A.10, A.11, and A.12.

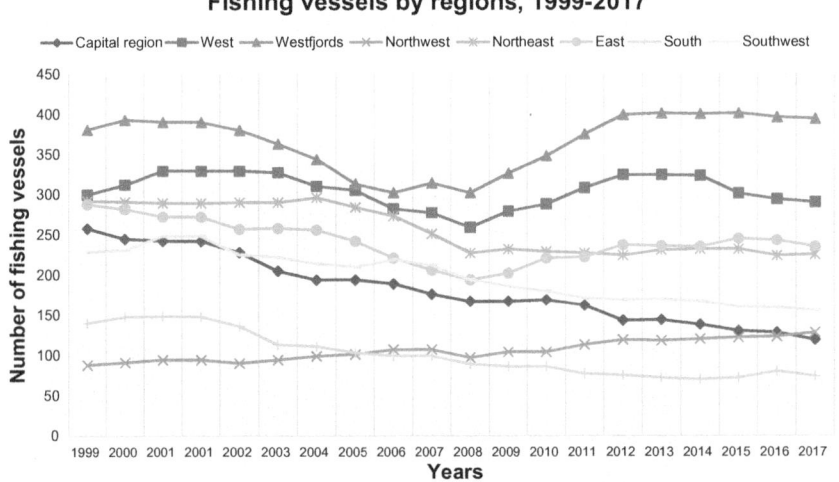

Fig. A.1 Fishing vessels by regions, 1999–2017. The graph clearly shows the Westfjords region holding the largest number of vessels (total 394 in 2017) of all regions in Iceland (including decked and undecked vessels and trawlers) (Statistics Iceland, 2018a). While vessel numbers in the Westfjords generally decreased with the consolidation of the quota system in the small boat system in the early 2000s, vessel numbers started rising again with the implementation of the summer coastal fisheries in 2009

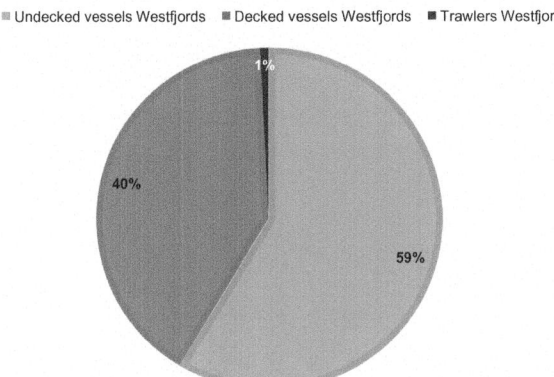

Fig. A.2 Fleet structure, Westfjords 2017. The fishing fleet consists of 231 undecked vessels (59%), a total of 160 decked vessels (40%) and only three trawlers (1%) (Statistics Iceland, 2018a)

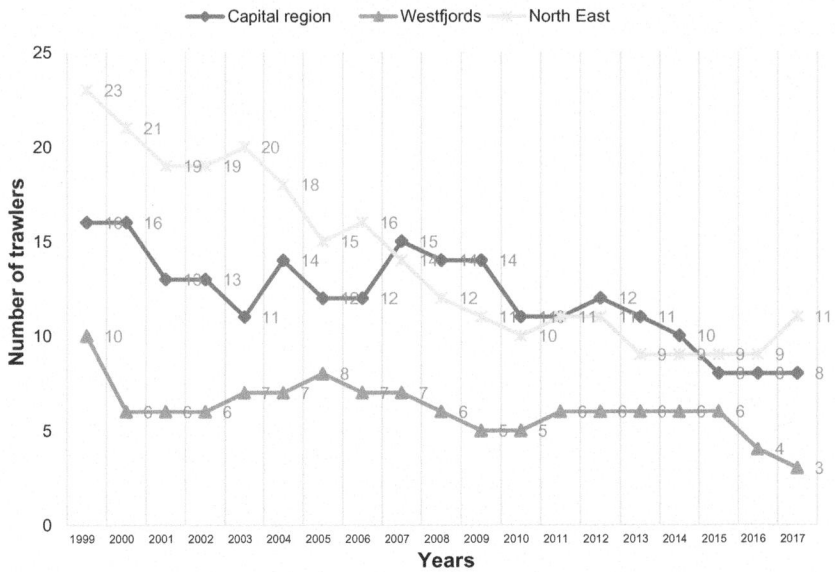

Fig. A.3 Trawlers by regions, 1999–2017. Although most fishing vessels are situated in the Westfjords, most of the trawler fleet is situated in the Capital region and the Northeast with Reykjavík and Akureyri as the most populated centres of the country (Statistics Iceland, 2018a)

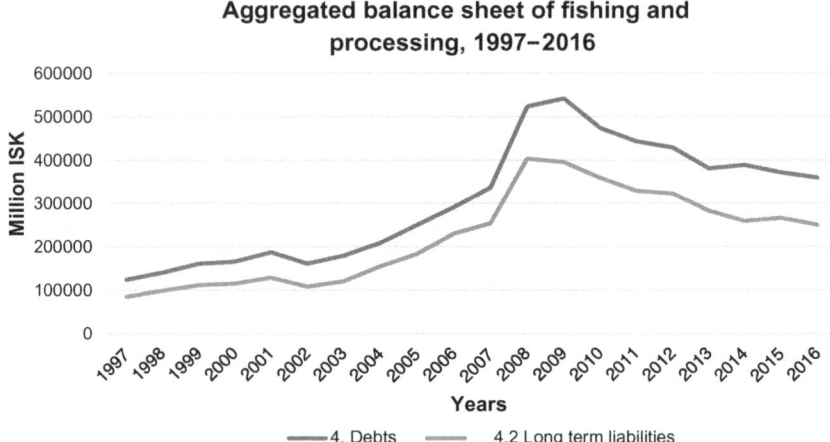

Fig. A.4 Debt and liabilities in fisheries and processing. The data suggests a steady increase from 39.368 to 54.1783 million ISK in debts, making for an increase of 1276.2% in the fishing industry, boosted by deregulation of the Icelandic economy and privatisation of banks in the early 2000s. A similar trend can be seen for long-term liabilities. The post-crisis development suggests a rather steep decrease, though debt and liabilities in the industry remain high (Statistics Iceland, 2018b)

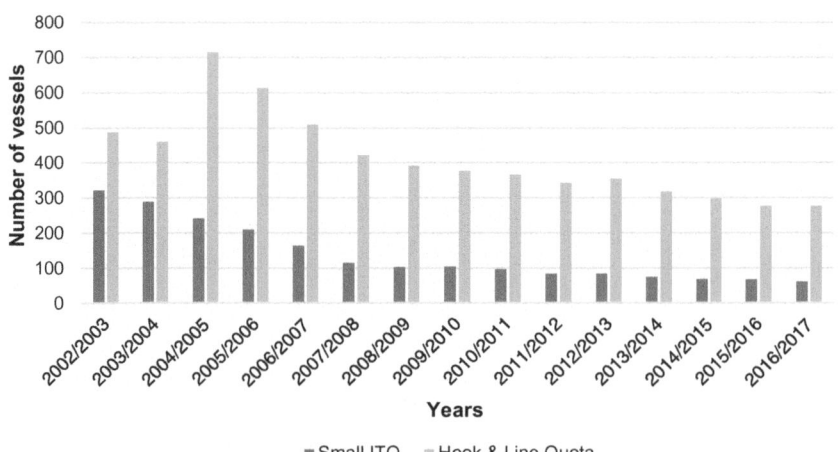

Fig. A.5 Small-boat fleet structure according to quota category, 2002–2018 (Fiskistofa, 2018)

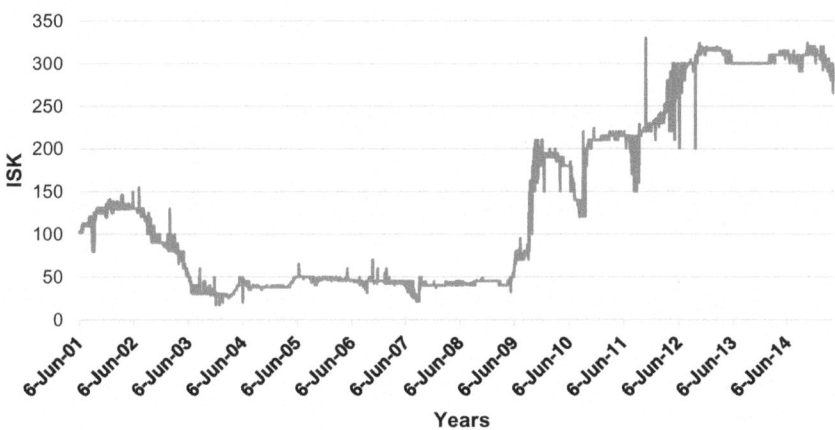

Fig. A.6 Rent prices haddock quota, 2001–2015. The data shows a steep and steady increase of the quota lease price for haddock from roughly 50 ISK in the period 2004–2008 to well above 300 ISK in 2012–2015, increasing the rent price over 500% (based on Fiskistofa, 2015)

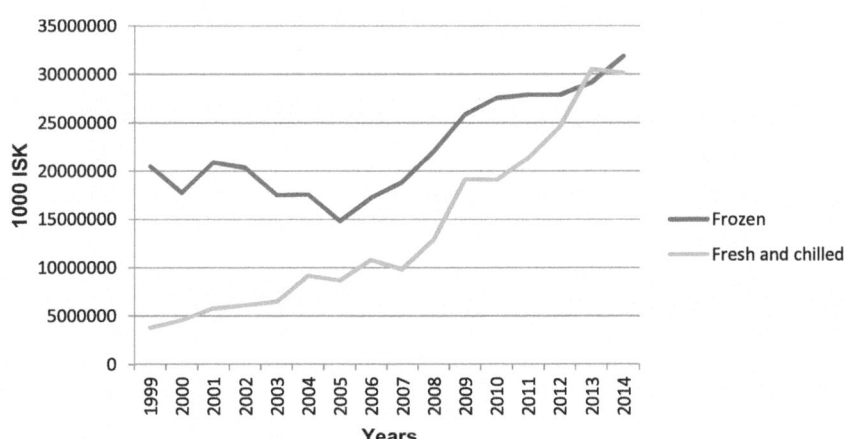

Fig. A.7 The development of salt- and fresh fish exports for cod, 1992–2014. The diagram illustrates the decline of processed salt fish and the back-to-front rise of fresh fish processing (Statistics Iceland, 2015b)

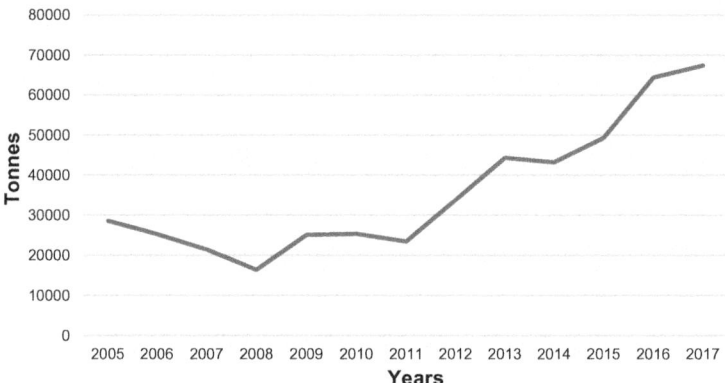

Fig. A.8 Fresh, iced fish exported by air, 2005–2017. According to the data pro-vided by Statistics Iceland (2015c), the air exports for both cod and haddock were below 5000 tonnes in 1992. From the mid-1990s, however, the data point to a slow but steadily increasing trend for both species, with cod peaking at 28,500 tonnes in 2005, then decreasing steadily until 2008. Since 2011, however, exports have increased steadily, peaking at 67,335 in 2017

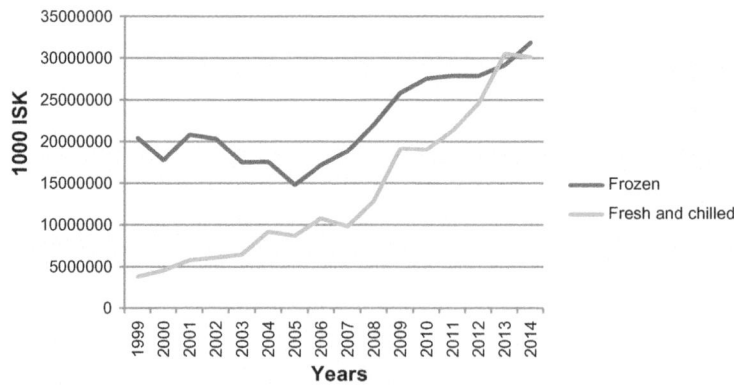

Fig. A.9 Value of marine products (FOB) (cod), 1999–2014. Since 1990, the value of fresh chilled cod has increased steadily, though even more rapidly during the financial expansion of the fishing industry in 2007. In 2013, the export value for fresh chilled cod even exceeded the export value for frozen and today remains high at over 30,000,000 thousand ISK per year (Statistics Iceland, 2015c)

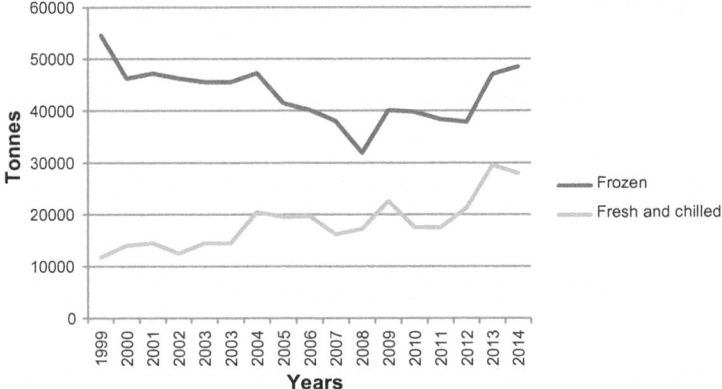

Fig. A.10 Quantities of marine products (cod), 1999–2014. Although the annual export value of frozen cod exports remains at a similar level to the annual export value for chilled fresh fish (see above), the quantities for chilled fish remain at a considerably lower level, with a difference of 20,528 tonnes in 2014 between the product categories (Statistics Iceland, 2015c)

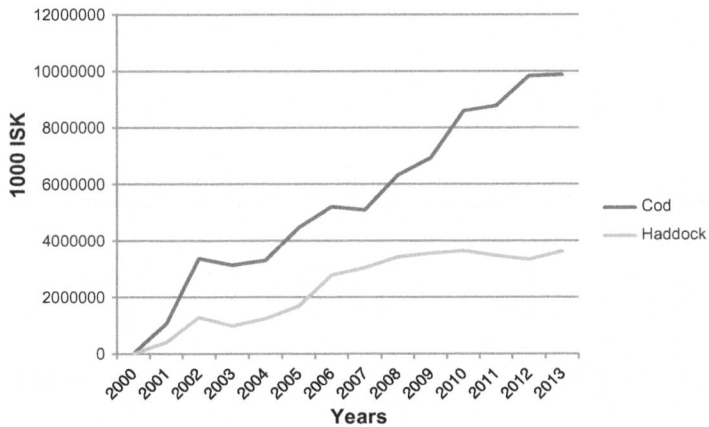

Fig. A.11 Value of catch in hook-and-line quota class. The graph illustrates the valorisation of hook-and-line landings since the year 2000 (no data are available prior to this year). Since 2007, the total catch value for cod has increased continuously to almost 99,000,000 thousand ISK. Note the increasing value of landings for haddock despite the steady cuts in haddock quotas, which have nevertheless increased the total value of raw materials to around 36,000,000 thousand ISK (Statistics Iceland, 2015a)

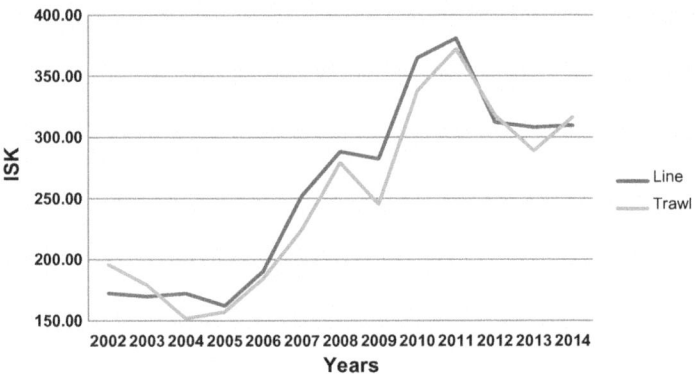

Fig. A.12 Annual average auction prices for line and trawl fish per kilo (cod), 2002–2014. Although auction prices for both trawl- and line-caught cod follow almost identical fluctuations and achieve the highest prices of all fishing gear used in the Icelandic fisheries, the graph shows a tendency to higher prices for line-caught fish from 2004 to 2011, leading to a difference of up to 37.19 ISK in 2009. In 2011, annual average prices for line fish peaked at a record average of 381.05 ISK (based on RSF, 2015; note that with 2012 the dataset distinguishes between 'line' and 'land baited line', with the latter being used for the analysis)

References

Fiskistofa. (2015). Leiguverð á ýsu í aflamarkskerfinu frá 6. júní 2001 til 14 apríl 2015. Retrieved from: https://www.google.com/url?sa=t&rct=j&q=&esrc=s&source=web&cd=1&ved=2ahUKEwji_qHmnYHeAhXCiywKHSxnDwYQFjAAegQI-CRAC&url=http%3A%2F%2Fwww.fiskistofa.is%2Fmedia%2Faflatolur%2F14042015_millifaerslur_dagverd_haestaverd_ysa_2001_2015.xls&usg=AOvVaw3FBBLGBHjO616euSXoNQ6m.

Fiskistofa. (2018). *Úthlutað aflamark fyrri ár.* Retrieved from http://www.fiskistofa.is/aflamarkheimildir/uthlutadaflamark/fyrriar/.

RSF. (2015). *Annual Average Auction Prices After Fishing Gear, 2002–2014.* Dataset received from Reiknistofu fiskmarkaða hf.

Sævaldsson, H., & Valtýsson, H. Þ. (2012). *Gillnets.* Retrieved from http://www.fisheries.is/fisheries/fishing-gear/gillnets/.

Sævaldsson, H., & Valtýsson, H. Þ. (2013). *Longline.* Retrieved from http://www.fisheries.is/fisheries/fishing-gear/longline/.

Statistics Iceland. (2011). *The Fishing Fleet at the End of 2010.* Retrieved July 20, 2015, from Statistics Icelandhttp://www.statice.is/lisalib/getfile.aspx?ItemID=12164.

Statistics Iceland. (2015a). *Catch and Value of Catch by Species, Quota Type and Months, 1992–2013.* Retrieved from http://px.hagstofa.is/pxen/pxweb/en/Atvinnuvegir/Atvinnuvegir__sjavarutvegur__aflatolur__kvotaflokkar/SJA09021.px.

Statistics Iceland. (2015b). *Catch of Icelandic Vessels by Type of Processing and Fishing Area.* Retrieved from http://www.statice.is/?PageID=1215&src=https://rannsokn.hagstofa.is/pxen/Dialog/varval.asp?ma=SJA09101%26ti=Catch+of+Icelandic+vessels+by+type+of+processing+and+fishing+area+1992-2013%26path=../Database/sjavarutvegur/radsFisk/%26lang=1%26units=Tons.

Statistics Iceland. (2015c). *Exported Marine Products by Product Categories and Species 1999–2014.* Retrieved from http://px.hagstofa.is/pxen/pxweb/en/Atvinnuvegir/Atvinnuvegir__sjavarutvegur__aflatolur__afli_verdmaeti/SJA02203.px.

Statistics Iceland. (2018a). *The Fishing Fleet by Region and Type of Vessel 1999–2017.* Retrieved from http://px.hagstofa.is/pxen/pxweb/en/Atvinnuvegir/Atvinnuvegir__sjavarutvegur__skip/SJA05001.px/?rxid=30a421c3-3c47-4f7d-8c70-a3ed851dca24.

Statistics Iceland. (2018b). *Aggregated Balance Sheet of Fishing and Fish Processing 1977–2018.* Retrieved from http://px.hagstofa.is/pxen/pxweb/en/Atvinnuvegir/Atvinnuvegir__sjavarutvegur__afkomasja/SJA08104.px/?rxid=9a3b3969-9bab-49ae-9ebf-bbd0cd0df082.

Þór, J. Þ. (2002). *Sjósókn og Sjávarfang. Saga Sjávarútvegs Á Íslandi.* Akureyri: Bókaútgáfan Hólar.

Þór, J. Þ. (2013). The History of Longline Fisheries in the Westfords Region. Personal Communication with Jón Þ. Þór.

Bibliography

Acheson, J. M. (1977). Technical Skills and Success in the Maine Lobster Industry. *Human Ecology, 3*(3), 183–207.

Acheson, J. M. (1981). Anthropology of Fishing. *Annual Reviews of Anthropology, 3*(3), 275–316.

Acheson, J. M. (1988). *The Lobster Gangs of Maine*. Hanover: University Press of New England.

Acheson, J. M. (1998). Lobster Trap Limits: A Solution to a Communal Action Problem. *Human Organization, 57*(1), 43–52.

Acheson, J. M. (2003). *Capturing the Commons: Devising Institutions to Manage the Main Lobster Industry*. Lebanon, NH: University Press of New England.

Ahrne, G., Aspers, P., & Brunsson, N. (2015). The Organization of Markets. *Organization Studies, 36*(1), 7–27.

Akerlof, G. A. (1970). The Market for Lemons: Uncertainty and the Market Mechanism. *The Quarterly Journal of Economics, 83*(3), 488–500.

Antal, A. B., Hutter, M., & Stark, D. (Eds.). (2015). *Moments of Valuation*. Oxford: Oxford University Press.

Apostle, R., Barret, G., Holm, P., Jentoft, S., Mazany, L., McCay, B., & Mikaelsen, K. (1998). *Community, State and Market on the North Atlantic Rim: Challenges to Modernity in the Fisheries*. Toronto: University of Toronto Press.

© The Editor(s) (if applicable) and The Author(s) 2019, corrected publication 2020 **243**
A. Dobeson, *Revaluing Coastal Fisheries*,
https://doi.org/10.1007/978-3-030-05087-0

Arnason, R. (2008). Iceland's ITQ System Creates New Wealth. *The Electronic Journal of Sustainable Development, 1*(2), 35–41.

Arnason, R., & Runolfsson, B. T. (Eds.). (2008). *Advances in Rights Based Fishing: Extending the Role of Property in Fisheries Management*. Reykjavík: Bókafélagið Ugla.

Aspers, P. (2009). *How Are Markets Made?* Retrieved from Köln.

Aspers, P. (2010a). *Orderly Fashion: A Sociology of Markets*. Princeton: Princetion University Press.

Aspers, P. (2010b). Using Design for Upgrading in the Fashion Industry. *Journal of Economic Geography, 10*(2), 189–207.

Aspers, P. (2011). *Markets*. Cambridge: Polity.

Aspers, P. (2018). Forms of Uncertainty Reduction: Decision, Valuation, and Contest. *Theory and Society, 2*(47), 133–149.

Barbera, F., & Audifredi, S. (2012). In Pursuit of Quality: The Institutional Change of Wine Production Market in Piedmont. *Sociologia Ruralis, 52*(3), 311–331.

Barnes, J. A. (1954). Class and Commitees in a Norwegian Island Parish. *Human Relations, 7,* 39–58.

Bazilchuk, N. (2010). Mackerel Wars. *Frontiers in Ecology and Environment, 8*(8), 397.

Bear, C., & Eden, S. (2008). Making Space for Fish: The Regional, Network and Fluid Spaces of Fisheries Certification. *Social & Cultural Geography, 9*(5), 487–504.

Beckert, J. (1996). What Is Sociological About Economic Sociology? Uncertainty and the Embeddedness of Economic Action. *Theory and Society, 25,* 803–840.

Beckert, J. (2009a). The Social Order of Markets. *Theory and Society, 38,* 245–269.

Beckert, J. (2009b). Wirtschaftssoziologie als Gesellschaftsstheorie. *Zeitschrift für Soziologie, 38*(3), 182–197.

Beckert, J. (2016). *Imagined Futures: Fictional Expectations and Capitalist Dynamics*. Harvard: Harvard University Press.

Beckert, J., & Aspers, P. (Eds.). (2011). *The Worth of Goods: Valuation & Pricing in the Economy*. Oxford: Oxford University Press.

Beckert, J., & Musselin, C. (2013a). Introduction. In J. Beckert & C. Musselin (Eds.), *Constructing Quality: The Classification of Goods in Markets* (pp. 1–23). Oxford: Oxford University Press.

Beckert, J., & Musselin, C. (Eds.). (2013b). *Constructing Quality: The Classification of Goods in Markets.* Oxford: Oxford University Press.

Benediktsson, K., & Karlsdóttir, A. (2011). Iceland: Crisis and Regional Development—Thanks for All the Fish? *European Urban and Regional Studies, 18*(2), 228–235.

Bestor, T. C. (2004). *Tsukiji: The Fish Market at the Center of the World.* Berkeley CA: University of California Press.

Bjarnason, T., & Thorlindson, T. (1993). In Defense of a Folk Model: The "Skipper Effect" in the Icelandic Cod Fishery. *American Anthropologist, 95*(2), 371–394.

Blattner, W. (2008). What Heidegger and Dewey Could Learn from Each Other. *Philosophical Topics, 36*(1), 57–77.

Boström, M., & Klintman, M. (2011). *Eco-Standards, Product Labelling and Green Consumerism.* Basingstoke: Palgrave Macmillan.

Bourdieu, P. (1977). *Outline of a Theory of Practice.* Cambridge: Cambridge University Press.

Braun, B. (2005). Environmental Issues: Writing a More-Than-Human Urban Geography. *Progress in Human Geography, 29*(5), 635–650.

Busch, L. (2007). Performing the Economy, Performing Science: From Neoclassical to Supply Chain Models in the Agrifood Sector. *Economy and Society, 36*(3), 437–466.

Busch, L. (2010). Can Fairy Tales Come True? The Surprising Story of Neoliberalism and World Agriculture. *Sociologia Ruralis, 50*(4), 331–351.

Byres, T. J. (1979). Of Neo-Populist Pipe-Dreams: Daedalus in the Third World and the Myth of Urban Bias. *The Journal of Peasant Studies, 6*(2), 210–244.

Çalişkan, K. (2007). Price as Market Device: Cotton Trading in Izmir Mercantile Exchange. In M. Callon, Y. Millo, & F. Muniesa (Eds.), *Market Devices* (pp. 241–260). Oxford: Blackwell.

Çalişkan, K. (2010). *Market Threads: How Cotton Farmers and Traders Create a Global Market.* Princeton: Princeton University Press.

Çalişkan, K., & Callon, M. (2009). Economization, Part 1: Shifting Attention from the Economy Towards Processes of Economization. *Economy and Society, 38*(3), 369–398.

Çalişkan, K., & Callon, M. (2010). Economization, Part 2: A Research Programme for the Study of Markets. *Economy and Society, 39*(1), 1–32.

Callon, M. (1986). Some Elements of a Sociology of Translation: Domestication of the Scallops and the Fishermen of St. Brieuc Bay.

In J. Law (Ed.), *Power, Action and Belief: A New Sociology of Knowledge?* (pp. 196–223). London: Routledge.

Callon, M. (1998a). An Essay on Framing and Overflowing. In M. Callon (Ed.), *The Laws of the Markets* (pp. 244–269). Oxford: Blackwell.

Callon, M. (1998b). *The Laws of the Market.* Oxford: Blackwell.

Callon, M. (1999). Actor-Network Theory—The Market Test. *The Sociological Review, 47*(1), 181–195.

Callon, M., Méadel, C., & Rabeharisoa, V. (2002). The Economy of Qualities. *Economy and Society, 31*(2), 194–217.

Callon, M., Millo, Y., & Muniesa, F. (Eds.). (2007). *Market Devices.* Malden: Blackwell.

Cardwell, E. (2015). Power and Performativity in the Creation of the UK Fishing-Rights Market. *Journal of Cultural Economy, 8*(6), 705–720.

Castells, M. (2001). *The Internet Galaxy: Reflections on the Internet, Business, and Society.* Oxford: Oxford University Press.

Chayanov, A. V. (1986). *The Theory of Peasant Economy.* Madison: The University of Wisconsin Press.

Chuenpagdee, R. (2011a). Too Big to Ignore: Global Research Network for the Future of Small-Scale Fisheries. In R. Chuenpagdee (Ed.), *World Small-Scale Fisheries: Contemporary Visions* (pp. 383–394). Delft: Eburon.

Chuenpagdee, R. (Ed.). (2011b). *World Small-Scale Fisheries: Contemporary Visions.* Delft: Eburon.

Cronon, W. (1991). *Nature's Metropolis: Chicago and the Great West.* New York: W. W. Nortan.

Dahrendorf, R. (1973). *Homo Sociologicus: Ein Versuch zur Geschichte, Bedeutung und Kritik der Kategorie der sozialen Rolle.* Opladen: Westdeutscher Verlag.

Davis, G. F. (2009). *Managed by the Markets: How Finance Reshaped America.* Oxford: Oxford University Press.

Demeter. (2015). *Fisch aus nachhaltiger Küstenfischerei.* Retrieved from http://www.felderzeugnisse.de/TK-Fisch.107.0.html.

Dequech, D. (2003). Uncertainty and Economic Sociology: A Preliminary Discussion. *American Journal of Economics and Sociology, 62*(3), 509–532.

Desmond, M. (2014). Relational Ethnography. *Theory and Society, 43,* 547–579.

Dobeson, A. (2016). Scopic Valuations: How Digital Tracking Technologies Shape Economic Value. *Economy and Society, 45,* 454–478.

Dobeson, A. (2018a). Between Openness and Closure: Helmuth Plessner and the Boundaries of Social Life. *Journal of Classical Sociology, 18*(1), 36–54.

Dobeson, A. (2018b). Economising the Rural: How New Markets and Technology Transform Rural Economies. *Sociologia Ruralis, 58*(4), 886–908.

Dobeson, A. (2018c). The Wrong Fish: Maneuvering the Boundaries of Market-Based Resource Management. *Journal of Cultural Economy, 11*(2), 110-124.

Dobeson, A. (2019, forthcoming). Das Fischerdorf im liberalen Kapitalismus: sozialräumliche Öffnungs- und Schließungsprozesse in der nordatlantischen Peripherie. In A. Steinführer, L. Laschewski, T. Mölders, & R. Siebert (Eds.), *Das Dorf. Soziale Prozesse und räumliche Arrangements*. Berlin: LIT.

Dreyfus, H. L. (1991). *Being-in-the-World: A Commentary on Heidegger's Being and Time, Division I*. New Baskerville: MIT Press.

Dreyfus, H. L. (1993). Heidegger's Critique of the Husserl/Searle Account of Intentionality. *Social Research, 60*(1), 17–38.

Dreyfus, H. L. (2004). Merlau-Ponty and Recent Cognitive Science. In C. Carman & M. Hansen (Eds.), *The Cambridge Companion to Merleau-Ponty* (pp. 129–150). Cambridge: Cambridge University Press.

Dreyfus, H. L. (2014). *Skillful Coping: Essays on the Phenomenology of Everyday Perception and Action*. Oxford: Oxford University Press.

Dreyfus, H. L., & Dreyfus, S. E. (1984). From Socrates to Expert Systems: The Limits of Calculative Rationality. *Technology in Society, 6*(3), 217–233.

Dziewicki, M. (2007). *The Role of AIS for Small Ships Monitoring*. Paper presented at the Baltic Master Workshop, Gdynia, 11–12 May 2006.

Einarsson, N. (2011a). *Culture, Conflict and Crises in the Icelandic Fisheries: An Anthropological Study of People, Policy and Marine Resources in the North Atlantic Arctic* (PhD dissertation). Uppsala University.

Einarsson, N. (2011b). Fisheries Governance and Social Discourse in Post-crisis Iceland: Responses to the UN Human Rights Committee's View in Case 1306/2004. *Yearbook of Polar Law, 3*, 470–515.

Einarsson, N. (2011c). Fisheries Governance and Social Discourse in Post-crisis Iceland: Responses to the UN Human Rights Committee's Views in Case 1306/2004. In N. Einarsson (Ed.), *Culture, Conflict and Crises in the Icelandic Fisheries: An Anthropoligical Study of People, Policy and Marine Resources in the North Atlantic Arctic*. Uppsala: Acta Universitatis Upsaliensis.

Esping-Andersen, G. (1985). *Politics Against Markets: The Social Democratic Road to Power*. Princeton: Princeton University Press.

Esping-Andersen, G. (1990). *Three Worlds of Welfare Capitalism*. Princeton: Princeton University Press.

European Commission. (2009). *The Common Fisheries Policy: A Users Guide.* Luxembourg: Office for Official Publications of the European Commission.

European Commission. (2012). *Control Technologies: The EU System for Fisheries Controls.* Retrieved from http://ec.europa.eu/fisheries/cfp/control/technologies/index_en.htm.

Eythórsson, E. (1996). Theory and Practice of ITQs in Iceland. *Marine Policy, 20*(3), 269–281.

Eythórsson, E. (2000). A Decade of ITQ-Management in Icelandic Fisheries: Consolidation Without Consensus. *Marine Policy, 24,* 483–492.

Fiskifréttir (Producer). (2017, 11 September 2018). *Mikilvægt að forðast ormaslóð.* Retrieved from http://www.fiskifrettir.is/frettir/mikilvaegt-ad-fordast-ormaslod/138149/.

Fiskistofa. (2006). *The Fisheries Management Act.* Reykjavík: Icelandic Ministry of Fisheries and Agriculture. Retrieved from http://www.fisheries.is/management/fisheries-management/the-fisheries-management-act/.

Fiskistofa. (2012). *Aflahlutdeild stærstu útgerðanna.* Retrieved from http://www.fiskistofa.is/umfiskistofu/frettir/nr/775.

Fiskistofa. (2013, September 26). *AIS Transponders in Iceland.* Personal Communication with Thorstein Hilmarsson.

Fiskistofa. (2015a). *Yfirlit yfir strandveiðar 2014.* Retrieved from http://www.fiskistofa.is/veidar/aflaupplysingar/yfirlit-sidasta-fiskveidiars/strandveidar/.

Fiskistofa. (2015b). *Total Catches of Species in the Icelandic Quota System.* Retrieved from http://www.fiskistofa.is/english/quotas-and-catches/total-catch-and-quota-status/?skipnr=0&timabil=0708&fyrirspurn=Um-Skip&landhelgi=i.

Fiskistofa. (2015c). Leiguverð á ýsu í aflamarkskerfinu frá 6. júní 2001 til 14. apríl 2015. Retrieved from https://www.google.com/url?sa=t&rct=j&q=&esrc=s&source=web&cd=1&ved=2ahUKEwji_qHmnYHeAhX-CiywKHSxnDwYQFjAAegQICRAC&url=http%3A%2F%2Fwww.fiskistofa.is%2Fmedia%2Faflatolur%2F14042015_millifaerslur_dagverd_haestaverd_ysa_2001_2015.xls&usg=AOvVaw3FBBLGBHjO616euSX-oNQ6m.

Fiskistofa. (2018). *Úthlutað aflamark fyrri ár.* Retrieved from http://www.fiskistofa.is/aflamarkheimildir/uthlutadaflamark/fyrriar/.

Fligstein, N. (1996). Markets as Politics: A Political-Cultural Approach to Market Institutions. *American Sociological Review, 61*(4), 656–673.

Fligstein, N. (2002). *The Architecture of Markets.* Princeton: Princeton University Press.

Foucault, M. (1970/2002). *The Order of Things: An Archeology of the Human Sciences*. London: Routledge.

Foucault, M. (1975). *Discipline and Punish: The Birth of the Prison*. New York: Vintage.

Foucault, M. (2008). *Security, Territory, Population: Lectures at the Collège de France 1978–1979*. Basingstoke: Palgrave Macmillan.

Foucault, M. (2009). *The Birth of Biopolitics: Lectures at the Collège de France 1978–1979*. Basingstoke: Palgrave Macmillan.

Friedmann, H., & McMichael, P. (1989). Agriculture and the State System: The Rise and Decline of National Agricultures, 1870 to the Present. *Sociologia Ruralis, XXIX*(2), 93–117.

Garcia-Parpet, M.-F. (2008). The Social Construction of a Perfect Market: The Strawberry Auction at Fontaines-en-Sologne. In *Do Economists Make Markets? On the Performativity of Economics* (pp. 20–53). Princeton: Princeton University Press.

Gatewood, J. B. (1984). Is the "Skipper Effect" Really a False Ideology? *American Ethnologist, 11*(2), 378–379.

Goffman, E. (1961a). Introduction. In E. Goffman (Ed.), *Asylums: Essays on the Social Situation of Mental Patients and Other Inmates* (pp. IX–XIV). New York: Random House.

Goffman, E. (1961b). *On the Characteristics of Total Institutions Asylums: Essays on the Social Situation of Mental Patients and Other Inmates* (pp. 3–124). New York: Random House.

Graham, I. (1998). The Emergence of Linked Fish Markets in Europe. *Electronic Markets, 8*(2), 29–33.

Graham, I. (1999). *The Construction of Electronic Markets* (PhD dissertation). University of Edinburgh, Edinburgh.

Granovetter, M. (1985). Economic Action and Social Structure: The Problem of Embeddedness. *American Journal of Sociology, 40*(3), 481–510.

Grundvåg, G. S., Larsen, T. A., & Young, J. A. (2013). The Value of Line-Caught and Other Attributes: An Exploration of Premiums for Chilled Fish in UK Supermarkets. *Marine Policy, 38*, 41–44.

Gulbrandsen, L. H. (2009). The Emergence and Effectiveness of the Marine Stewardship Council. *Marine Policy, 33*, 654–660.

Habermas, J. (1981). *Theorie des kommunikativen Handelns. Band 2. Zur Kritik der funktionalistischen Vernunft*. Frankfurt am Main: Suhrkamp.

Hall, P. A., & Soskice, D. (2001). *Varieties of Capitalism: The Institutional Foundations of Comparative Advantage*. Oxford: Oxford University Press.

Halldórsson, G. H. (2010). *Strandveiðarnar 2009: Markmið, framgangur og fiskveiðistjórnun* (Master's thesis). University of Akureyri, University Centre of the Westfjords, Ísafjörður. Retrieved from https://skemman.is/handle/1946/5668.

Hamsun, K. (1921). *Growth of the Soil*. New York: A. A. Kopf.

Hannesson, R. (1991). From Common Fish to Rights Based Fishing: Fisheries Management and the Evolution of Exclusive Rights to Fish. *European Economic Review, 35*, 397–407.

Hannesson, R. (2004). *The Privatization of the Oceans*. Cambridge: MIT Press.

Haraway, D. J. (1985). A Manifesto for Cyborgs: Science, Technology and Socialist Feminism in the 1980s. *Socialist Review, 80*, 65–107.

Hardin, G. (1968). The Tragedy of the Commons. *Science, 162*, 1243–1248.

Harris, J. (Ed.). (1982). *Rural Development: Theories of Peasant Economy and Agrarian Change*. London: Hutchnis University Library.

Harvey, D. (2005). *A Brief History of Neoliberalism*. Oxford: Oxford University Press.

Heidegger, M. (1962). *Being and Time*. New York City: Harper Collins.

Heidegger, M. (1977). The Question Concerning Technology. In M. Heidegger (Ed.), *The Question Concerning Technology* (pp. 3–35). New York City: Harper Collins.

Heidegger, M. (1983). Gesamtausgabe. II. Abteilung: Vorlseungen 1923–1944. Band 29/30. In *Die Grundbegriffe der Metaphysik. Welt - Endlichkeit - Einsamkeit*. Frankfurt am Main: Vittorio Klostermann.

Helgason, A., & Pálsson, G. (1997). Contested Commodities: The Moral Landscape of Modernist Regimes. *The Journal of the Royal Anthropological Institute, 3*(3), 451–471.

Hersoug, B. (1997). What is Good for the Fishermen, is Good for the Nation: Co-management in the Norwegian Fishing Industry in the 1990s. *Ocean & Coastal Management, 35*(2–33), 157–172.

Hersoug, B. (2002). *Unfinished Business: New Zealand's Experience with Rights-Based Fisheries Management*. Delft: Eburon.

Hersoug, B. (2005). *Closing the Commons: Norwegian Fisheries from Open Access to Private Property*. Delft: Eburon.

Hersoug, B., Holm, P., & Rånes, S. A. (2000). The Missing T. Path Dependency Within an Individual Vessel Quota System—The Case of the Norwegian Cod Fisheries. *Marine Policy, 24*, 319–330.

Hirschman, A. O. (1982). Rival Interpretations of Market Society: Civilizing, Destructive, or Feeble? *Journal of Economic Literature, 20*, 1463–1484.

Hollan, J., Hutchins, E., & Kirsh, D. (2000). Distributed Cognition: Toward a New Foundation for Human-Computer Interaction Research. *ACM Transactions on Computer-Human Interaction, 7*(2), 174–196.

Hollingsworth, R. J., & Boyer, R. (Eds.). (1997). *Contemporary Capitalism: The Embeddedness of Institutions.* Cambridge: Cambridge University Press.

Holm, P. (1995). The Dynamics of Institutionalization: The Transformation Process in Norwegian Fisheries. *Administrative Science Quarterly, 40*(3), 398–422.

Holm, P. (2001). *The Invisible Revolution: The Construction of Institutional Change in the Fisheries* (PhD dissertation). University of Tromsø, Tromsø.

Holm, P. (2007). Which Way Is Up on Callon? In F. Muniesa, L. Siu, & D. MacKenzie (Eds.), *Do Economists Make Markets?* (pp. 225–243). Princeton: Princeton Univeristy Press.

Holm, P., & Nolde Nielsen, K. (2007). Framing Fish, Making Markets: The Construction of Individual Transferable Quotas. In Y. Millo, M. Callon, & F. Muniesa (Eds.), *Market Devices* (Vol. 55, pp. 173–195). Malden: Blackwell.

Høst, J. (2015). *Market-Based Fisheries Management: Private Fish and Captains of Finance.* Dordrecht: Springer.

Hutchins, E. (1995a). *Cognition in the Wild.* Cambridge, MA: MIT Press.

Hutchins, E. (1995b). How a Cockpit Remembers Its Speeds. *Cognitive Science, 19*, 265–288.

Icelandic. (2015). Products & Markets: Fresh Fish. Retrieved from http://www.icelandic.is/icelandic/products-markets/fresh-fish/.

Ingimundarson. (2008). Fighting the Cod Wars in the Cold War: Iceland's Challenge to the Western Alliance in the 1970s. *The RUSI Journal, 148*(3), 88–94.

Jentoft, S. (1989). Fisheries Co-management. *Marine Policy, 13*(2), 137–154.

Joas, H. (1992). *Die Kreativität des Handelns.* Frankfurt am Main: Suhrkamp.

Johnsen, J. P. (2004). The Evolution of the "Harvest Machinery": Why Capture Capacity Has Continued to Expand in Norwegian Fisheries. *Marine Policy, 29*, 481–493.

Johnsen, J. P. (2013). Is Fisheries Governance Possible? *Fish and Fisheries, 15*(3), 428–444.

Johnsen, J. P., Holm, P., Sinclair, P., & Bavington, D. (2009). The Cyborgization of the Fisheries: On Attempts to Make Fisheries Management Possible. *Mast, 7*(2), 9–34.

Johnsen, J. P., Murray, G. D., & Neis, B. (2009). North Atlantic Fisheries in Change: From Organic Associations to Cybernetic Organizations. *Mast, 7*(2), 55–82.

Karpik, L. (2010). *Valuing the Unique: The Economics of Singularities*. Princeton: Princeton University Press.

Keohane, N. O., & Olmstead, S. (2007). *Markets and the Environment*. Washington, DC: Island Press.

Knight, F. H. (1921). *Risk, Uncertainty and Profit*. Boston: Houghton Mifflin.

Knorr Cetina, K. (1989). Spielarten des Konstruktivismus. Einige Notizen und Anmerkungen. *Soziale Welt, 40*(1/2), 86–96.

Knorr Cetina, K. (1999). *Epistemic Cultures: How the Sciences Make Knowledge*. Cambridge, MA: Harvard University Press.

Knorr Cetina, K. (2001). Objectual Practice. In T. R. Schatzki, K. Knorr Cetina, & E. v. Savigny (Eds.), *The Practice Turn in Contemporary Theory* (pp. 176–188). London and New York: Routledge.

Knorr Cetina, K. (2003). From Pipes to Scopes: The Flow Architecture of Financial Markets. *Distinktion: Scandinavian Journal of Social Theory, 4*(2), 7–23.

Knorr Cetina, K. (2009). The Synthetic Situation: Interactionism for a Global World. *Symbolic Interaction, 32*(1), 61–87.

Knorr Cetina, K., & Bruegger, U. (2002). Global Microstructures: The Virtual Societies of Financial Markets. *American Journal of Sociology, 107*(4), 905–950.

Knorr Cetina, K., & Preda, A. (Eds.). (2005). *The Sociology of Financial Markets*. Oxford: Oxford University Press.

Lamont, M. (2012). Toward a Comparative Sociology of Valuation and Evaluation. *Annual Review of Sociology, 38*, 201–221.

Latour, B. (1999). *Pandora's Hope: Essays on the Reality of Science Studies*. Cambridge, MA: Harvard University Press.

Latour, B. (2005). *Reassembling the Social: An Introduction to Actor-Network-Theory*. Oxford: Oxford University Press.

Law, J., & Hassard, J. (1999). *Actor Network Theory and After*. Oxford: Blackwell.

Laxness, H. (2008). *Independent People*. London: Vintage.

Lemke, T. (2001). The Birth of Bio-Politics: Michel Foucault's Lecture at the Collège de France on Neo-Liberal Governmentality. *Economy and Society, 30*, 190–207.

Luhmann, N. (1993). *Risk: A Sociological Theory*. Berlin and New York: Walter de Gruyter.

Mackenzie, A. F. (2006). A Working Land: Crofting Communities, Place and the Politics of the Possible in Post-Land Reform Scotland. *Transactions of the Institute of British Geographers, 31*(3), 383–398.

MacKenzie, D. (1984). Marx and the Machine. *Technology and Culture, 25*(3), 473–502.

MacKenzie, D. (2009a). Constructing Emission Markets. In D. MacKenzie (Ed.), *Material Markets: How Economic Agents Are Constructed* (pp. 137–176). Oxford: Oxford University Press.

MacKenzie, D. (2009b). *Material Markets.* Oxford: Oxford University Press.

MacKenzie, D., & Wajcman, J. (1999a). Introductory Essay: The Social Shaping of Technology. In D. MacKenzie & J. Wajcman (Eds.), *The Social Shaping of Technology.* Buckingham: Open University Press.

MacKenzie, D., & Wajcman, J. (Eds.). (1999b). *The Social Shaping of Technology* (2nd ed.). Buckingham: Open University Press.

Macneil, I. (1978). Contracts: Adjustment of Long-Term Economic Relations Under Classical, Neoclassical and Relational Contract Law. *Northwestern University Law Review, 72,* 854–887.

Marinetraffic.com. (2013). *Frequently Asked Questions.* Retrieved from http://www.marinetraffic.com/ais/de/faq.aspx?level1=160#2.

MAST. (2012). *Hitastigsmælingar á lönduðum afla.* Retrieved from Selfoss.

Matís. (2010). *Mikilvægi góðrar meðhöndlunar á fiski.* Retrieved from Reykjavík: http://www.matis.is/media/matis/utgafa/Mikilvaegi-godrar-medhondlunar-a-fiski.pdf.

McMichael, P. (2005). Global Development and the Corporate Food Regime. In P. M. Frederick & H. Buttel (Eds.), *New Directions in the Sociology of Global Development* (Vol. 11, pp. 269–303). Bingley: Emerald.

Merleau-Ponty, M. (2012/1945). *Phenomenology of Perception.* Abingdon, Oxon: Routledge.

Miller, M., & Van Maanen, J. (1982). Getting into Fishing: Observations on the Social Identities of New England Fishermen. *Journal of Contemporary Ethnography, 11*(1), 27–54.

Miller, M. L., & Van Maanen, J. (1979). "Boats Don't Fish, People Do": Some Ethnographic Notes on the Federal Management of Fisheries in Gloucester. *Human Organization, 38*(4), 377–385.

Münch, R. (2012). *Inclusion and Exclusion in the Liberal Competition State: The Cult of the Individual.* London and New York: Routledge.

Muniesa, F., Millo, Y., & Callon, M. (2007). An Introduction into Market Devices. In M. Callon, Y. Millo, & F. Muniesa (Eds.), *Market Devices* (pp. 1–11). Oxford: Blackwell.

Norris, A. (2007). AIS Implementation—Success or Failure? *The Journal of Navigation, 60*(1), 1–10.

Ostrom, E. (1990). *Governing the Commons: The Evolution of Institutions for Collective Action.* Cambridge: Cambridge University Press.

Pálsson, G. (1991). *Coastal Economies, Cultural Accounts: Human Ecology and Icelandic Discourse*. Manchester: Manchester University Press.

Pálsson, G. (1994). Enskilment at Sea. *Man, 29*(4), 901–927.

Pálsson, G., & Durrenberger, E. P. (1982). To Dream of Fish: The Causes of Icelandic Skipper's Fishing Success. *Journal of Anthropological Research, 38*(2), 227–242.

Pálsson, G., & Durrenberger, E. P. (1990). Systems of Production and Social Discourse: The Skipper Effect Revisited. *American Anthropologist, 92*(1), 130–141.

Pálsson, G., & Helgason, A. (1995). Figuring Fish and Measuring Men: The Individual Transferable Quota System in the Icelandic Cod Fishery. *Ocean & Coastal Management, 28*(1–3), 117–146.

Pálsson, G., & Helgason, A. (1996). The Politics of Production: Enclosure, Equity, and Efficiency. In G. Pálsson & E. P. Durrenberger (Eds.), *Images of Contemporary Iceland: Everyday Lives and Global Contexts* (pp. 60–86). Iowa City: University of Iowa Press.

Parsons, T. (1951). *The Social System*. Glencoe: The Free Press.

Pechlaner, G., & Otero, G. (2010). The Neoliberal Food Regime: Neoregulation and the New Division of Labor in North America. *Rural Sociology, 75*(2), 179–208.

Phillipson, J., & Symes, D. (2015). Finding a Middle Way to Develop Europe's Fisheries Dependent Areas: The Role of Fisheries Local Action Groups. *Sociologia Ruralis, 55*(3), 343–359.

Polanyi, K. (1957). The Economy as Instituted Process. In M. Granovetter & R. Swedberg (Eds.), *The Sociology of Economic Life* (pp. 31–50). Cambridge: Westview Press.

Polanyi, K. (2001). *The Great Transformation: The Political and Economic Origins of Our Time*. Boston: Beacon.

Reckwitz, A. (2002). Toward a Theory of Social Practices: A Development in Culturalist Theorizing. *European Journal of Social Theory, 5*(2), 243-263.

Rees, J. (2013). *Refrigeration Nation: A History of Ice, Appliances, and Enterprise in America*. Baltimore: John Hopkins.

RSF. (2015a). *Annual Average Auction Prices After Fishing Gea, 2002–2014*. Dataset received from Reiknistofu fiskmarkaða hf.

RSF. (2015b). *Haddock Auction Market Price, 2012–2014*. Dataset received from RSF. Dataset received from Reiknistofu fiskmarkaða hf.

Sævaldsson, H., & Valtýsson, H. Þ. (2012). *Gillnets*. Retrieved from http://www.fisheries.is/fisheries/fishing-gear/gillnets/.

Sævaldsson, H., & Valtýsson, H. Þ. (2013). *Longline*. Retrieved from http://www.fisheries.is/fisheries/fishing-gear/longline/.

Sandberg, C. (2014). *On Board the Global Workplace* (PhD dissertation). Stockholm University, Stockholm.

Schatzki, T. R. (1996). *Social Practices: A Wittgensteinian Approach to Human Activity and the Social*. Cambridge: Cambridge University Press.

Schatzki, T. R. (2001). Practice Mind-ed Orders. In T. R. Schatzki, K. Knorr Cetina, & E. v. Savigny (Eds.), *The Practice Turn in Contemporary Theory* (pp. 42–55). London and New York: Routledge.

Schatzki, T. R., Knorr Cetina, K., & Savigny, E. v. (2001). *The Practice Turn in Contemporary Theory*. London and New York: Routledge.

Schimank, U. (2008, February). Kapitalistische Gesellschaft - differenzierungstheoretisch konzipiert. Beitrag zur Tagung der Sektion Wirtschaftssoziologie der Deutschen Gesellschaft für Soziologie "Theoretische Ansätze der Wirtschaftssoziologie", Berlin.

Schimank, U. (2014). Modernity as Functionally Differentiated Capitalist Society: A General Theoretical Model. *European Journal of Social Theory, 18*(4), 413–430.

Schimank, U., & Volkman, U. (Eds.). (2012). *The Marketization of Society: Economizing the Non-Economic*. Bremen: University of Bremen.

Schumacher, E. F. (1973). *Small Is Beautiful: A Study of Economics as If People Mattered*. London: Blond & Briggs.

Schütz, A., & Luckmann, T. (2003/1975). *Strukturen der Lebenswelt*. Konstanz: UVK Verlagsgesellschaft/ UTB.

Shucksmith, M., & Brown, D. L. (2016). Framing Rural Studies in the Global North. In M. Shucksmith & D. L. Brown (Eds.), *Routledge International Handbook of Rural Studies* (pp. 1–26). London and New York: Routledge.

Shucksmith, M., & Rønningen, K. (2011). The Uplands After Neoliberalism?—The Role of the Small Farm in Rural Sustainability. *Journal of Rural Studies, 27*, 275–287.

Skaptadóttir, U. D. (1996). Housework and Wage Work: Gender in Icelandic Fishing Communities. In G. Pálsson & E. P. Durrenberger (Eds.), *Images of Contemporary Iceland: Everyday Lives and Global Contexts* (pp. 87–105). Iowa City: University of Iowa Press.

Standal, D., & Aarset, B. (2002). The Tragedy of Soft Choices: Capacity Accumulation and Lopsided Allocation in the Norwegian Coastal Cod Fishery. *Marine Policy, 26*, 221–230.

Stark, D. (2009). *The Sense of Dissonance: Accounts of Worth in Economic Life*. Princeton: Princeton University Press.

Statistics Iceland. (2013a). *Population by Locality, Age and Sex 1 January 2011–2013.* Ísafjörður.

Statistics Iceland. (2013b). *Population by Municipalities, Sex and Age 1 January 1998–2013—Current Municipalities.* Ísafjarðarbær.

Statistics Iceland. (2015a). *Catch and Value of Catch by Species, Quota Type and Months, 1992–2013.* Retrieved from http://px.hagstofa.is/pxen/pxweb/en/Atvinnuvegir/Atvinnuvegir__sjavarutvegur__aflatolur__kvotaflokkar/SJA09021.px.

Statistics Iceland. (2015b). *Catch of Icelandic Vessels by Type of Processing and Fishing Area.* Retrieved from http://www.statice.is/?PageID=1215&src=https://rannsokn.hagstofa.is/pxen/Dialog/varval.asp?ma=SJA09101%26ti=Catch+of+Icelandic+vessels+by+type+of+processing+and+fishing+area+1992-2013%26path=../Database/sjavarutvegur/radsFisk/%26lang=1%26units=Tons.

Statistics Iceland. (2015c). *Exported Marine Products by Product Categories and Species 1999–2014.* Retrieved from http://px.hagstofa.is/pxen/pxweb/en/Atvinnuvegir/Atvinnuvegir__sjavarutvegur__aflatolur__afli_verdmaeti/SJA02203.px.

Statistics Iceland. (2017). *Population by Municipalities, Sex and Age 1 January 1998–2017—Current Municipalities.* Westfjords.

Statistics Iceland. (2018a). *The Fishing Fleet by Region and Type of Vessel 1999–2017.* Retrieved from http://px.hagstofa.is/pxen/pxweb/en/Atvinnuvegir/Atvinnuvegir__sjavarutvegur__skip/SJA05001.px/?rxid=30a421c3-3c47-4f7d-8c70-a3ed851dca24.

Statistics Iceland. (2018b). *Aggregated Balance Sheet of Fishing and Fish Processing 1977–2018.* Retrieved from http://px.hagstofa.is/pxen/pxweb/en/Atvinnuvegir/Atvinnuvegir__sjavarutvegur__afkomasja/SJA08104.px/?rxid=9a3b3969-9bab-49ae-9ebf-bbd0cd0df082.

Strauss, A. L. (1978). *Negotiations: Varieties, Contexts, Processes, and Social Order.* San Francisco: Jossey-Bass.

Svallfors, S., & Tyllström, A. (2018). Resilient Privatization: The Puzzling Case of for-Profit Welfare Providers in Sweden. *Socio-Economic Review.* Retrieved from https://doi.org/10.1093/ser/mwy005.

Sverisson, Á. (2002). Small Boats and Large Ships: Social Continuity and Technical Change in the Icelandic Fisheries, 1800–1960. *Technology and Culture, 43*(2), 227–253.

Tetreault, B. J. (2005). *Use of the Automatic Identification System (AIS) for Maritime Domain Awareness (MDA).* Paper presented at the Oceans, Washington, DC.

Thévenot, R. (1979). *A History of Refrigeration Throughout the World.* Paris: International Institute of Refrigeration.

van de Walle, G., da Silva, S. G., O'Hara, E., & Soto, P. (2015). Achieving Sustainable Development of Local Fishing Interests: The Case of Pays d'Auray Flag. *Sociologia Ruralis, 55*(3), 360–377.

Velthuis, O. (2005). *Talking Prices: Symbolic Meanings of Prices on the Market for Contemporary Art.* Princeton: Princeton University Press.

Wagner, P. (1994). *A Sociology of Modernity: Liberty and Discipline.* London: Routledge.

Whatmore, S. (2006). Materialist Returns: Practicing Cultural Geography in an for a More-Than-Human World. *Cultural Geographies, 13*(4), 600–609.

White, H. C. (1981). Where Do Markets Come From? *The American Journal of Sociology, 87*(3), 517–547.

White, H. C. (2002). *Markets from Networks: Socioeconomic Models of Production.* Princeton: Princeton University Press.

White, H. C. (2008). *Identity and Control: How Social Formations Emerge.* Princeton: Princeton University Press.

Williamson, O. (1975). *Markets and Hierarchies: Analysis and Anti-Trust Implications: A Study in the Economics of Internal Organization.* New York: Free Press.

Willson, M. (2016). *Seawomen of Iceland: Survival on the Edge.* Seattle and London: University of Washington Press.

Þór, J. Þ. (2002). *Sjósókn og Sjávarfang. Saga Sjávarútvegs Á Íslandi.* Akureyri: Bókaútgáfan Hólar.

Þór, J. Þ. (2013, November 5). The History of Longline Fisheries in the Westfords Region. Personal Communication with Jón Þ. Þór.

Þórðarson, G., & Viðarsson, J. R. (2014). *Coastal Fisheries in Iceland.* Retrieved from http://www.matis.is/media/matis/utgafa/12-14Coastal-fisheries-in-Iceland.pdf.

Index

© The Editor(s) (if applicable) and The Author(s) 2019, corrected publication 2020
A. Dobeson, *Revaluing Coastal Fisheries*,
https://doi.org/10.1007/978-3-030-05087-0

The manufacturer's authorised representative in the EU is Springer
Nature Customer Service Centre GmbH, Europaplatz 3, 69115 Heidelberg,
Germany. If you have any concerns regarding our products, please
contact ProductSafety@springernature.com

Printed and bound by CPI Group (UK) Ltd, Croydon, CR0 4YY
29/04/2026
02099450-0006